RAGNAROKING!

RAGNARÖK, COLLAPSE, ODINISM, & MATHEMATICS

VIKING AGE BARBARIAN

DISCLAIMER

This work is an artistic work of satire in the Swiftian/Juvenalian tradition, published for entertainment and information purposes only. Neither the publisher, nor author assumes any responsibility for the use of misuse of this book's information. Readers undertaking weapons and self-defense training need to be aware of the laws of their jurisdiction. Before undertaking any exercise program, obtain an examination and clearance from a medical doctor, a proctologist if necessary. No-one should want to die before the fun starts at Ragnarök.

The book is solely concerned with personal defense in a collapse of civilization situation, which means the end of organized and orderly society regulated by social institutions and encompasses the WROL – Without Rule of Law scenario – meaning that "legitimate" or conventional government has ceased to exist, and the rule of law cannot function, so that there are no laws to break, and no legitimate body enforcing the law through the courts. Assertions in this book relate to such a WROL situation, and not to present civil, or uncivil, society, and even here in a literary kind of way.

978-0-6487660-8-7
Ragnaroking! Ragnarök, Collapse, Odinism, & Mathematics.
Viking Age Barbarian.

© Manticore Press, Melbourne, Australia, 2021.

Thema Classification:
QRSW (Norse religion), VXWS (Paganism), VSY (Survivalism), JWXZ (Self-Defence), PB (Mathematics).

MANTICORE PRESS
WWW.MANTICORE.PRESS

CONTENTS

INTRODUCTION

Into the Wastelands

Neo-barbarian fans of *Bronze Age Mindset* (Kindle, 2018), by the great Bronze Age Pervert, a nude body builder, may enjoy this book, which goes even further in its iconoclasm, embracing the Stone Age Mindset, of hyper-barbarianism, but as I like Viking weapons, and that is my tribe, here I am. Further, while embracing the combat arts and power training, especially training in the wild lifting huge rocks and logs and throwing them, doing this in the nude is not advocated. Have you ever had a bee sting your long and dangly? Or, have you charged naked through a thorn bush? Who will remove the thorns embedded in your meat after the collapse of civilization? Food for perverts, indeed.

This postmodern iconoclastic version of *Aesop's Fables*, sans the cuddly creatures, is the ultimate toilet read for the intellectually constipated. It consists of iconoclastic essay after essay, influenced by fellow neo-barbarian, James LaFond, in an unrelenting volley of anti-establishment thought… I hope. The LaFond site, and the works of that homeless, crazed martial artist, machete and stick fighting badass, fit into a perennially pissed off dissident paradigm that sees modern techno-industrial civilization as, to use a technical term, "fucked,"[1] facing impending collapse, or it may be dead already,

---

[1]   See Brad Werner, "Is Earth Fucked?" Paper delivered to the American Geophysical Union, 2012, http://www.declineoftheempire.com/2012/12/is-the-earth-fcked.html; S. Motesharrei (et al.), "Human and Nature Dynamics (HANDY): Modeling Inequality and

moving on in zombie-like motions. Others may piss themselves in fear, urine running down their legs, pooling in a nervous puddle at the thought of the final TEOTWAWKI, but this wolf age of Ragnarök, or *Kali Yuga*, for fans of the Sanskrit scriptures, is embraced with cackling joy by LaFond and this present nihilistic crank; hence the present title: *Ragnaroking!* ... sardonic glee at the world's end, accelerationism on steroids and growth hormone. Why? Because for many of us, it is better that the cesspool world ends than it continues, its said zombified existence being explained in another mighty Manticore Press book, Thorfinn Skullsplitter, *Zombie Apocalypse Now!* (July, 2019), which almost out-sold my magnificent, *The Barbarian Reborn* (2020), not that I am the jealous type. We will take our chances on the apocalypse, as the present fetid shithole is unliveable, castrating, crippling, and existentially nauseating, and that's on a good day. Few have put the anticipated joy for the new barbarian more succinctly than fellow iconoclast Corey Savage, author of *Man's Fight for Existence*, (Kindle, 2016), in a piece, "4 Reasons Why Collapse Will be the Best Thing to Happen for Men," (June 20, 2016), celebrating the return of the primal order:

> The collapse will mean the restoration of natural order: the rule of the jungle. In fact, I think it would be wrong to call the destruction of our so-called civilization as a "collapse"; it would simply be a return to the way things were. No more corporate serfdom, no more putrid consumerism, no more technological slavery, and no safe spaces for the cry-babies to hide and cry in. Wimps, complainers, and the weak will not survive. People will once again be naturally selected instead of being artificially sheltered.

> One of the best aspects of the new order would be the return of masculine virtue. As I've said, any new society that people form must be defended against external threats. This is not an option. And only an organized group of men with strength, courage,

---

Use of Resources in the Collapse or Sustainability of Societies," *Ecological Economics*, vol. 101, 2014, pp. 90-102.

mastery, and honor (as per Jack Donovan) will prevail in the post-apocalyptic world. Men will be men again.

Who knows what savage energy is begging to be unleashed within that man serving as an office drone? Who knows if that guy flipping burgers for a minimum wage will become the future tribal leader? How many men today are living jaded and unfulfilling lives when they could be fighters and warriors instead?

Can't wait until men are allowed to be men again? Then you better be ready for the war for survival.[2]

The West is caught and being filleted like a wriggling fish by the dark serrated blade of the elites and their weapons of mass social destruction. But, impending destruction and the doom of modernity, is not all overall bad, since collapse, as we will see, has its upside. Thus, like Ragnar Loðbrók, we laugh as it dies, and if we die.[3] This is the end, but we intend to go, if necessary, by the Donovanian way of men, holding to the warrior virtues as we perish, as the best, and most complete Nietzschean beasts we can be, before adding to the rotting carcass of this fetid Earth, joyful corpses.[4]

Predictions of the collapse of the West and with it techno-industrial society are far from new, but what is new today is the massive body of evidence of crises in everything, peak everything,[5]

---

[2]   https://www.returnofkings.com/88671/4-reasons-why-collapse-will-be-the-best-thing-to-happen-for-men.

[3]   "It gladdens me to know that Baldr's father (Odin) makes ready the benches for a banquet. Soon we shall be drinking ale from the curved horns. The champion who comes into Odin's dwelling (Valhalla) does not lament his death. I shall not enter his hall with words of fear upon my lips. The Æsir will welcome me. Death comes without lamenting. Eager am I to depart. The Dísir summon me home, those whom Odin sends for me (Valkyries) from the halls of the Lord of Hosts. Gladly shall I drink ale in the high-seat with the Æsir. The days of my life are ended. I laugh as I die." "The Death Song of Ragnar Lodbrok": http://www.englishmonarchs.co.uk/vikings_9.html.

[4]   J. Donovan, *The Way of Men*, (Kindle edition, 2012), *Becoming a Barbarian*, (Kindle edition, 2016), *A More Complete Beast*, (Kindle Edition, 2018).

[5]   R. Heinberg, *Peak Everything*, (New Society Publishers, Kindle edition, 2010), *The End*

from the global financial system imploding, to local environmental destruction and the collapse of global ecosystem, at least as it is capable of sustaining human life.[6] Problems of resource depletion, and radical changes to complex, unpredictable chaotic non-linear systems, such as the world climatic system, have an enormous scientific and technical literature, generally ignored by conservatives, being in general ignorant fuckwits.[7] However, it does not matter what horseman of the apocalypse you may be skeptical about, such as peak oil or climate change, because there are plenty of others to take their place, such as biodiversity destruction, looming food shortages, the population bomb, ethnic warfare and much more. How many times over do we need to die?

Popular expressions of the significance of this convergence of compounding, interacting, converging catastrophes, have been made by thinkers from both the Left and the Right. Thus, we have the popular survey from the Left by John Michael Greer, *Dark Age America: Climate Change, Cultural Collapse, and the Hard Future Ahead*,[8] seeing industrial society grinding to a halt and the world entering a new Dark Age, which will, at least for some time, be violent, as people will not go easily and quietly into the darkest of nights. A similar position was taken by that other intelligent man of the Left, James Howard Kunstler in *The Long Emergency*, also arguing that humanity is facing a set of converging catastrophes.[9] A good neutral book on the coming economic/ecological collapse is Piero San Giorgio, *Survive the Economic Collapse*,[10] with my own *The*

---

of Growth, (Kindle edition, 2012).

[6]   N. M. Ahmed, *A User's Guide to the Crisis of Civilization: and How to Save It*, (Pluto Press, London, 2010).

[7]   For a summary for the general reader, taking a biohistorical, evolutionary perspective, see: E. Kolbert, *The Sixth Extinction: An Unnatural History*, (Bloomsbury, London, 2015).

[8]   John Michael Greer, *Dark Age America: Climate Change, Cultural Collapse, and the Hard Future Ahead*, (New Society Publishers, Kindle edition, 2016).

[9]   James Howard Kunstler, *The Long Emergency: Surviving the End of Oil, Climate Change, and Other Converging Catastrophes of the Twenty-First Century*, (Atlantic Monthly Press, New York, 2005); http://kunstler.com/.

[10]   Piero San Giorgio, *Survive: The Economic Collapse*, (Radix/Washington Summit,

Barbarian Reborn, being utterly outstanding, and one book every home should have, more user friendly than king sized condoms, with raised "ribs" for her pleasure.

In the Right corner, there are numerous works proclaiming our doom, including a now classic by the late Pentti Linkola (1932-2020), *Can Life Prevail?*[11] but an outstanding comprehensive one, although now a bit dated on the technical data, is by the late Guillaume Faye (1949-2019), *Convergence of Catastrophes*,[12] which is largely in agreement with Greer and Kunstler on the environmental crisis, seeing the terrestrial ecosystem as collapsing ("it is already too late").[13] That is rare, because the Right, would in general rather have group anal with rhinoceroses than contemplate that there could be evidence (outside of the NWO/IPCC) of ecological damage. The God of capitalism gave us a right to mine life itself, and fuck it over they shall. For example, because of a drop-off in fertility in the West, thanks to the joys of women's liberation conservatives say, the diabolical economic elites want open borders and unlimited immigration to keep up profits, seemingly believing that taking in the surplus population of Africa, with its population explosion, will do the trick. How could this possibly go wrong?

Faye, as a man of the Right, argues that social chaos and breakdown will occur in Western societies, especially Europe, as a result of open borders migration. His book was written long before the so-called European "migration crisis" really got going, and lengthy books have been filled, and numerous articles written, describing the socio-political significance of the numerous rapes and violence, murders, and beheadings, often involving knives and machetes, that have followed in a return already to a state resembling the new Dark Ages, or at least techno-feudalism. Various internet sites cover this material daily, usually with a link to a mainstream news site, with

---

Whitefish, 2013), with a foreword by James Howard Kunstler, is one of the most comprehensive, easy to read overviews of the forces that will produce the coming collapse.

[11]   P. Linkola, *Can Life Prevail?* (Arktos, London, 2011).

[12]   Guillaume Faye, *Convergence of Catastrophes*, (Arktos, London, 2011).

[13]   As above, p. 26.

each day bringing a new shocker, pushing the boundaries of gothic horror. By way of a quickie summary: the grooming rapes of perhaps one million British children and thousands more in most European countries; ghoulish knife and machete murders; dismembered bodies and beheadings with body parts scattered untidily over the landscape; sexual slavery and more even more beheadings; Sweden, diverse assault rifles and hand grenades; impending European demographic chaos; Africa's population explosion, and South Africa's dark night. If this is not "filleting," even with some metaphorical flourish, then it will do until the main "deadly serious," game begins.

As Will Durant has said (quoted in the opening of Mel Gibson's film *Apocalypto* (2006)) "A great civilization is not conquered from without until it destroys itself from within." The combination of the converging and compounding environmental and social crises pose a super wicked problem that defies ready political solution, having been left to fester for so long. Even perhaps slightly less intense problems, such as the feminist attack on manhood, have been regarded by critics such as Camille Paglia, herself a "sensible feminist," as capable on its own of wrecking civilization, by "poisoning" it, and destroying manhood, and Samson-like, bringing down the pillars of the temple of decomposing modernity.

Then, there are the events of June 2020, following on from the "virus we cannot name in print" fiasco, "discussed" by fellow author Professor Thomas Rivers in *Invisible Enemies: Disease, Social Collapse, and Survivalism,* using "[…]" as place holder for coronavirus. We have seen crazed ferals burning down businesses, the killing of a young mother for merely saying that "all lives matter;" statues being pulled down or burnt as sacrificial effigies (destruction of statues is always flowed by destruction of people); anti-gunners marching with semi-auto rifles; areas of cities being seized and made into new states immediately ruled by warlords with illegal guns, but soon collapsing, and much more.

If 2020 has proven anything it is that conventional politics is in ashes.

This turn of events has led to many conservative types making proclamations that Viking Age Barbarian would make. The United States is irredeemably corrupt. It cannot be salvaged and it cannot be saved. The entire political and economic infrastructure is lost. America is now a failed state. "We are indeed, "riding a tiger," and it is best not to make oneself a tasty meal for the tiger, but to allow the beast to consume our enemies instead. It may sound heartless, but take it from an old guy. Once you get your affairs in order, sit back, relax, and enjoy the carnage … this is going to be one apocalypse you won't want to miss." That is the true spirit of *Ragnaroking*; head butting the chaos.

The essays to follow proceed on the basis that techno-industrial civilization is doomed, as humanity has been too smart for its own good,[14] and in particular, Western society, its history, culture and peoples are being fully cancelled,[15] and the cancellers are going to get canceled too, in the Great Cancellation, of every fucking thing. As stated, our crisis is multidimensional, comprising compounding and synergistically interacting components that feed off of each other, including the environmental crisis, the political crisis and economic crisis, and ethno-social crises, detailed in the literature referenced above.[16] This breakdown in the fundamental social and life support systems, will push an already decaying system over the abyss, and it is surprising that it has not happened sooner. But, happen it will.

What then? Clearly the time ahead will be one of mega-violence, as even some academics such as Peter Turchin have predicted.[17] Already, over a third of Americans think that civil war is likely,

---

[14] C. Dilworth, *Too Smart for Our Own Good: The Ecological Predicament of Humankind*, (Cambridge University Press, Cambridge, 2010).

[15] https://www.washingtonexaminer.com/opinion/editorials/yes-cancel-culture-is-real-and-its-dangerous.

[16] See https://survivalskills.guide/responding-to-collapse-adaptation-and-resilience-amidst-the-decline-of-civilization/.

[17] P. Turchin, *Ages of Discord: A Structural-Demographic Analysis of American History*, (Beresta Books, Chaplin, 2016); M. Papenfuss, "Society Could Collapse in a Decade, Predicts Math Historian," January 7, 2017, at http://www.huffingtonpost.com.au/entry/peter-turchin-cliodynamics-society-collapse_us_586f1e22e4b02b5f85882988.

and that could become a self-fulfilling prophecy.[18] Some of the consequences of "Amerika" becoming a "failed state" have been detailed by Health Ranger Mike Adams:

Failed infrastructure (power grid, telecommunications, water supplies, etc.).

- Total collapse of the rule of law (anarchy, chaos, cops quitting en masse, etc.).

- Financial collapse, followed by an explosion of nationwide riots, during which most U.S. cities will be gutted.

- Executions and kidnappings.

- Bank failures and the collapse of financial institutions.

- Political anarchy, attempted revolutions, assassinations of key leaders, etc.

- The rise of private security contractors to protect businesses with armed, military-trained guards.

- A worsening of drug abuse and suicides.

- Increases in child trafficking and child kidnappings for the trafficking trade.

- Rapidly escalating censorship, including browser-based blocking of targeted websites.

- An explosion in homelessness and tent cities as destitution and despair spreads across the nation.

- Civil war and secession as the nation fragments into smaller nation-states.[19]

As I see it, that is the optimistic prediction. One problem among many is that of the high levels of mental illness, if not outright batshit craziness associated with those presently taking jack hammers to the foundations of an already decaying West, let alone our deranged

---

[18]   https://www.zerohedge.com/political/over-third-americans-think-civil-war-likely.

[19]   https://yournews.com/2020/06/21/1693072/america-is-now-a-failed-state-no-rule-of-law/;https://www.jameslafond.com/article.php?id=12315.

ruling elites, most of whom should be in mental institutions, with heavy drugs pumped through their veins. But the lunatics now run the asylum that is modern society.[20] According to the Pew Research Center, as observed by New York journalist Lance Welton, 45.9 percent of "Liberals" aged between 18 and 29 have been diagnosed with a mental illness, compared to 20 percent of "Conservatives".[21] And as Welton has also pointed out, evidence indicates that many mental illnesses can spread as a form of social contagion, depressing birth rates, and undermining Western societies, producing a cult of destruction, which we now witness in its early stages. That alone, apart from every other perverse diverse and vibrating "ism," is sufficient to poison the West. The seeds of the tree have been irreparably contaminated, as surely as if one left one's genitals in a dental X-ray machine for a century.

How has all of this occurred, this pathology on a poison drip? The basic idea here is that primarily Northern Europeans (Nordics/ Nordish), behave as if they are like insects that are manipulated by "zombie parasites," who cause them to act contrary to their own group interests and survival.[22] Why?

Professor Kevin MacDonald hypothesizes that Northern Europeans, because of challenges posed by a harsh climate in pre-history, developed a strong individualism and low group solidarity (why not high, as this would have a selective advantage?), and while this had an alleged evolutionary advantage in the past, today, in contest with much more collectivistic, tribalist groups, they are systematically dispossessed, and ultimately demographically replaced. MacDonald's *Individualism and the Western Liberal Tradition: Evolutionary Origins, History, and Prospects for the Future* (2019), details a biohistorical and genetically based evolutionary

---

[20]   https://vdare.com/articles/it-s-official-again-social-justice-warriors-are-mentally-ill-and-they-re-contagious; E. J. Dawydiak (et al.), "Pathogen Disgust Predicts Stigmatization of Individuals with Mental Health Conditions," *Evolutionary Psychological Science*, vol. 6, 2020, pp. 60-63.

[21]   As above.

[22]   http://news.nationalgeographic.com/news/2014/10/141031-zombies-parasites-animals-science-halloween/.

history of Europeans, with a focus upon the individualism of this tragicomic sub-race of humanity. He traces the formation of Europeans from three groups: (1) Western Hunter Gatherers; (2) Indo-Europeans, and (3) Early Farmers. All of these groups intermixed to a degree. However, Northern Europeans are primarily composed of Western Hunter Gatherers, Southern Europeans by Early Farmers. Indo-Europeans conquered Europe 4,500 years ago and became its aristocratic rulers. The Western Hunter Gatherers correspond, approximately, to the Nordics of William Z. Ripley, Madison Grant and Carleton Coon, and they embraced a culture of egalitarian individualism, although still forming homogeneous moral communities. By contrast, the Early Farmers, a more collectivistic group, originated in Anatolia and brought with them agriculture.

The Indo-Europeans, arose in an area north of the Black Sea, formed from the populations of the Caucasian Hunter Gatherers, Ancient North Eurasians and Eastern Hunter Gatherers. They are also called Yamnayas. These muscular horse riders of doom spread across Europe, slaughtering men and impregnating their lamenting women, wiping out Neolithic tribesmen. They reached Britain 4,400 years ago, and eliminated the Britons who built Stonehenge.[23] They interbred with the Western Hunter Gatherers, and Early Farmers, but primarily the Western Hunter Gatherers. This Indo-European mixture was the driving force of Europeans, forming an aristocratic elite based upon aristocratic individualism. MacDonald sees a historical shift after the English Civil War (1642-1651) towards the egalitarian individualism of the Western Hunter Gatherers. The Puritan Revolution destroyed the Indo-European order based upon a military aristocracy. From that point on, egalitarian individualism was ascendant, and emerged victorious in other conflicts, such as the American Civil War. This is the source of out-group infiltration: liberal and egalitarian values.

Viking Age Barbarian appreciates this philosophical anthropology, but sees some limitations as an explanation of the pathological altruism seen today. First, the high altitude and harsh

---

[23]   https://www.dailymail.co.uk/sciencetech/article-6865741/the-violent-group-people-lived.html.

climate thesis fails, because other peoples in the same climate, or harsher, such as the Inuit, did not develop pathological altruism. And, note that it is pathological *altruism*, not pathological *individualism* that is supposed to account for today's racially suicidal behavior in the West. The harsh climate encountered in pre-history, would seem to favor tribalism over go-it-alone individualism, since it would take a high degree of social cooperation to arrange hunting and all other tasks as such in the lands of ice and snow, with loner individuals getting eaten by wolves. Thus, from an evolutionary perspective, we would expect the opposite from his hypothesis of individualism from the ice and snow dwellers; not individualism, but altruism, or at least a high degree of social cooperation.

The concept of pathological altruism is closely related to Garrett Hardin's notion of "promiscuous altruism."[24] In the abstract to his classic paper he stated:

> Reliable Darwinian theory shows that pure altruism cannot persist and expand over time. All higher organisms show inheritable patterns of caring and discrimination. The principal forms of discriminating altruisms among human beings are individualism (different from egoism), familialism, cronyism, tribalism, and patriotism. The promiscuous altruism called "universalism" cannot endure in the face of inescapable competition. Information can be promiscuously shared, but not so matter and energy without evoking the tragedy of the commons. Universalism is not recommended even as an ideal. Survival now requires the creation of an intellectual base for a new patriotism.

Hardin is correct in pointing his finger at the doctrine of *universalism* as the key culprit, standing behind individualism and egalitarian as responsible, not only for the once greatness of the West, the Faustian ideal and notions of manifest destiny, but also for the present-day downfall of the West, as such an idea becomes over-drawn and pathological.

---

24 G. Hardin, "Discriminating Altruisms," *Zygon*, vol. 17, 1982, pp. 163-186.

Thus, in the pre-World War II era, all of the settler nations of the US, Canada, Australia and New Zealand were liberal states with a sense of collective identity and homogeneity, which was attacked in the post-world War II period, by various elites. All of the arguments for deconstructing these societies were based ultimately upon universalist doctrines, even if made by groups with their own tribal interests, arguments such as non-discrimination, human rights and associated concepts. Universalism even was used to give birth to multiculturalism. As I argued in *The Barbarian Reborn*, universalism is internally incoherent.

Universalism is likely to be a part of the cognitive psychology of the Western mind, or of "WEIRD" people, to use Joseph Heinrich's term, Western, Educated, Industrialized, Rich and Democratic.[25] Psychological studies based upon the characteristics of WEIRD people, a thin slice of humanity, are likely to be unrepresentative of humanity in general, regarding many aspects of perception, decision-making, spatial cognition and other psychological characteristics. The Western analytic mind is more concerned with differences between things, than similarities, and WEIRD individuals see themselves as individuals first, before being seen as members of an in-group. The source of universalism lies in this cognitive state, especially in the need for on-going rational justifications of things rather than mere conventional acceptance by way of tradition. This rationalism, aligned with the quest to be universal, has formed Western philosophy, theories of truth, logic, mathematics, and moral and political philosophies. Thus, in ethics, the universalizability principle of Kantian style deontological ethics, provides that for acts to be permissible they must be applicable to all people. These universal principles, arising in Greek philosophy, in Christianity, and through to the Enlightenment, with its rational basis of egalitarianism, are the founding basis of the West, while individualism is secondary. It is entirely conceivable that the West could have evolved with a rugged Nietzschean/John Wayne individualism, but with ethnocentric relativism, rather than

---

[25]   J. Heinrich, *The Weirdest People in the World: How the West became Psychologically Peculiar and Particularly Prosperous*, (Penguin, New York, 2020).

universalism, particularly if Christianity did not get its dark talons into Nordic man.

Christianity, as many have observed, and as I argued in detail in *The Barbarian Reborn*, has been an example of the metaphysics of universalism from its beginning, as Pierre Krebs in *Fighting for the Essence* (2002) (and Dr Tomislav Sunic in many works), have also noted.[26] Thus, in my less than humble opinion, Professor Andrew Fraser in *Dissent Dispatches* (2017), is mistaken in hoping to see WASPs, as a Nordic ethnic group, "regenerate a collective spirit of in-group solidarity" from a return to a reformed Christianity.[27] Amalric de Droevig has made the same suggestion more recently, but also admits that contemporary Christianity has become highly toxic, and to be accepted it must

> abandon many of its universalistic aspects. It must emphasize the local, the hierarchical, and the traditional. It must deemphasize the global, the novel, and the egalitarian … Frankly, I don't see that happening, but time will tell.[28]

And, as Philip Jenkins shows in *The Next Christendom* (2011), Christianity is well on the way to becoming a religion of the Third World, returning to its origin.[29] But, as I argued in *The Barbarian Reborn*, and as Ferdinand Bardamu has also put it:

> Christianity is a perversion of the instinct for self-preservation. This makes it a destroyer of entire civilizations and people. Embracing Christianity is an act of suicide for those who allow

---

[26]   P. Krebs, *Fighting for the Essence: Western Ethnosuicide or European Renaissance*, (Arktos, London, 2002), p. 13.

[27]   A. Fraser, *Dissent Dispatches: An Alt Right Guide to Christian Theology*, (Arktos, London, 2017), p. 107.

[28]   See https://www.theoccidentalobserver.net/2020/07/18/the-way-forward-partition-partition-partition-and-the-remainder-of-the-operational-plan/.

[29]   P. Jenkins, *The Next Christendom: The Coming of Global Christianity*, (Oxford University Press, Oxford, 2011).

themselves to be influenced by its doctrines.[30]

Christianity also magnifies individualism and egalitarianism, and that alone makes it a problem.

Odinism, defined as "the organic spiritual beliefs and religion of the indigenous peoples of Northern Europe as embodied in the Eddas [and Sagas] and as they have found expression in the wisdom and historical experiences of these peoples,"[31] offers one alternative for the Nordic tribe, especially in the post-apocalyptic world, where the present-day religions of materialism and scientific rationalism have collapsed, as detailed in *The Barbarian Reborn*. Less extreme versions of Odinism, not linked to collapseology, include Pete Jennings, *The Norse Tradition* (1998); Osred, *Odinism: Present, Past and Future* (2010); Colin Cleary, *Summoning the Gods* (2011) and Stephen McNallen, *Asatru: A Native European Spirituality* (2015). However, collapseology does answer a core question posed by Colin Cleary in *Summoning the Gods*, of how from today's disadvantage point, the old gods are met? Well, the vacuum created by the loss of the World-As-We-Knew-It, will need to be filled, and the tribal remnant will need a social glue that a particularistic religion such as Odinism will readily supply. And, it does not matter if it is "true" or not in a sense that analytical philosophers define.

Some may propose that even in the post-apocalyptic ruins, it is still preferable to abandon all gods, seek to become a Nietzschean overman, becoming a god oneself, the ultimate expression of the Faustian quest. And, I have sympathies for this as detailed in *The Barbarian Reborn*. But, few will be able to pursue that path, so a tribal religion will be needed. Either way, one must become a barbarian, as I detailed in *The Barbarian Reborn*, casting off the universalistic trappings of modernity. It is lost anyway, so move on.

My Odinism is basically a "Thorism," Thor not depicted in the *Eddas*, but the earlier pre-Odinic God, with no written records,

---

[30]   F. Bardamu, "Why Europeans Must Reject Christianity: Part I," April 3, 2018, at https://www.eurocanadian.ca/2018/04/why-europeans-must-reject-christianity.html.

[31]   Osred, *Odinism: Present, Past and Future*, (Renewal Publications, Melbourne, 2010), p. 5.

but a record burnt into our soul.[32] Further, I reject the idea that the Norse Gods were polished off at Ragnarök, a literary device where the snotty Christian scribes conveniently put an end to our Gods, having the liberty to write history. Thus, the account in the *Eddas* where Thor dies from poison from the Midgard Serpent is rejected, since other sources hold that Thor had earlier killed the Serpent, so he would have been free to help Odin defeat the Fenris Wolf. As observed by Lotte Motz:

> Þórr's weapon was forged for him in the smithy of some dwarfs to serve as a missile and as a weapon of close attack (*Skáldskaparmál* ch. 35). It would never fail, no matter how hard the blow, and it would return to the owner of its own accord when it was cast. Þórr indeed threw the hammer in his duel with the giant Hrungnir, and he broke the giant's head into small bits: hann ... reiddi hamarinn ok kastaði um langa leið at Hrungni (he ... swung his hammer and threw it from a great distance at Hrungnir; *Skáldskaparmál* ch. 17). He flung his weapon also at the Midgard snake, as he was fishing in the ocean, and it is said that the monster's head was struck from the body: Þórr kastaði hamrinum eptir honum, ok segja menn at hann lysti af honum höfuðit vid grunninum (Þórr threw his hammer after it, and they say that he struck off its head by the sea-bed; *Gylfaginning* ch. 48). In the Eddic poem which relates the same event, the head was merely battered by the tool before the fishing line was cut (Hymisqviða st. 23; *Edda* 1983, 92):

> Hamri kníði háfiall carar,
> ofliótt, ofan úlfs hnitbróður.

> With his hammer he struck down upon the most ugly head (hair's high mountain)
> of the wolf's inseparable (or battle-) brother.

---

32   C. Coon, *The Races of Europe*, (Macmillan, New York, 1939), p. 322.

A skaldic poem, Úlfr Uggason's Húsdrápa, also tells the story; here the head, hewn from the body, was sent into the sea:

Víðgymnir laust Vimrar
vaðs af fránum naðri
hlusta grunn við hrönnum.

Víðgymnir of Vimur's ford struck the head (ear-bed) from the shining
snake by the waves (*Skáldskaparmál* ch. 4).[33]

Ragnarök is for us; it is the twilight of humanity, our ending, the ending of modern civilization, and descent into a new Dark Age, not the destruction of the Norse gods, who fight their own battles against the Dark Gods of Relentless Evil, whose human reps rule us at present.

Whatever philosophy is adopted, it is evident that humans today are living under the artificial conditions parallel to the "mouse utopia" experiment of John B. Calhoun, where a mouse population in a "utopia" of adequate material comfort, and no struggle for existence and natural selection, ultimately imploded.[34] The contemporary team lead by the wonderfully named Michael A. Woodley of Menie, sees the same collapse of human civilization occurring as genetic mutations increase, intelligence decreases and natural selection is artificially suppressed by the "social trap" of civilization. The collapse will mean though that natural selection returns with a vengeance, flushing away all of this universalistic nonsense down nature's sewer, whether Christian or liberal.[35]

---

[33]   Lotte Motz, "The Germanic Thunder Weapon," *Saga-Book*, vol. 24, part 5, 1997, pp. 329-350, at p. 329

[34]   J. B. Calhoun, "Death Squared: The Explosive Growth and Demise of a Mouse Population," *Proceedings of the Royal Society of Medicine*, vol. 66, 1973, pp. 80-88.

[35]   M. A. Woodley of Menie (et al.), "Social Epistasis Amplifies the Fitness Costs of Deleterious Mutations, Engendering Rapid Fitness Decline Among Modernized Populations," *Evolutionary Psychological Science*, vol. 3, 2017, pp. 181-191; E. Dutton and M. Woodley of Menie, *At Our Wit's End: Why We are Becoming Less Intelligent and What it*

Thus, one has a choice of either being prey, or struggling to survive in the battle for life, as the ship of postmodernity sinks. The Norse Viking spirit, embraced here and in *The Barbarian Reborn*, is that of inner strength and lawful resistance, while the rule of law exists, even though one perishes. Valhalla awaits ... maybe ...

To cash out as a neo-Viking, rather than to die on one's knees like a battery hen or hog, requires a spiritual transformation, embracing one's inner barbarian. To do so the dominant oppressive ideologies must be cast aside. There has been more than enough critique of liberalism, the philosophy of the "hollow men" (T. S. Eliot), the grey NPCers that control the nerve centers of society. The first block of essays here goes back to where I think the rot all started, with Christianity. The arguments here are directed against the standard conservative narrative that Christianity constitutes the core heritage of the West. Yes, certainly there has been a strong cultural and ideological influence, but it is not as great as conservatives think, with the Nordic West changing Christianity as much as it changed the West.[36] Further, there is no reason to suppose that the might and greatness of the West,[37] or what it once was, was causally due to Christianity; *after* does not necessarily imply *because of*, the *post hoc, ergo propter hoc* fallacy. Christianity was the foul wind, against which the kite of the West rose. To vary the metaphor, today, it joins with its bastard child, malignant liberalism,[38] spawn from its savage womb,[39] as one of the forces working to destroy the tribe of Odin/

Means for the Future, (Imprint Academic, Exeter, 2018); J. Platt, "Social Traps," *American Psychologist*, vol. 28, 1973, pp. 641-651. Lance Welton has also reached this same conclusion, seeing the collapse of civilization and the return of natural selection: https://vdare.com/articles/is-antifa-a-spiteful-mutation-are-we-headed-to-a-mouse-utopia-collapse.

[36]   J. C. Russell, *The Germanization of Early Medieval Christianity: A Sociohistorical Approach to Religious Transformation*, (Oxford University Press, Oxford, 1996).

[37]   L. Brown, *The Might of the West*, (Robert B. Luce, 1979).

[38]   R. H. Bork, *Slouching Towards Gomorrah: Modern Liberalism and American Decline*, (Harper Perennial, New York, 2003).

[39]   L. Siedentop, *Inventing the Individual: The Origins of Western Liberalism*, (Belknap Press, New Haven, 2014); S. McLennar, *Jesus was a Liberal*, (Palgrave Macmillan, Hampshire, 2009).

Thor. Christianity has thus been a slow virus of the mind.[40] The essays in the first section oppose this, not to save the world we have and are about to lose, but to free the minds of the barbarians to come (if they need it), who will "wander the post-apocalyptic wastelands," doing the badass things which barbarians will do, when the world as we know it collapses. And, consistent with the anti-intellectualism of this neo-Viking barbarian, the essays are for fighters, not academics, whom I despise.

As a caveat: *this is not an attack upon Christian persons, but rather the doctrine of Christianity*. Further, although there is mention of white nationalism in this book, this is by way of criticism; we all need to go beyond nationalism and all modernistic doctrines, and embrace tribalism within a cultural relativist position, as Jack Donovan and others have convincingly argued in my opinion.[41] Hence, this book is more tolerant than conservative literature, which is unashamedly universalistic following the Enlightenment tradition, even accepting that some cultures may prefer witchcraft and sorcery to Western materialism. It is not a tribesman's role to convince outsiders of anything, only to live in peace away from them, until the world falls apart; then almost everyone dies.™ [42]

The second block of essays deal with subjects based around the ideal of being a barbarian, casting off the values and constraints of civilization and letting our savage meat swing free. The idea is to show how the neo-barbarian of the near future should, and will think. As these essays were written on a day-by-day basis; there is no systematic treatise, but more a blow-by-blow response to various topics.

Apart from philosophical speculation about the way of the barbarian, there are essays on melee weapons and their use. Sure, Americans have guns, but who knows for how long, given

---

[40]   R. Brodie, *Virus of the Mind*, (Hay House, Carlsbad, 2011).

[41]   https://www.jack-donovan.com/axis/2017/05/why-i-am-not-a-white-nationalist/;
Frank Salter, *On Genetic Interests: Family, Ethnicity, and Humanity in an Age of Mass Migration*, (Routledge, New York, 2017).

[42]   D. Consiglio, *Spoiler Alert: Everyone Dies,*™ (BEG Publishing, 2017).

the events of 2020. Further, other jurisdictions severely control personal firearm ownership, so melee weapons are as relevant as ever, especially as the filleting of the West goes on in the post-apocalyptic wastelands to come. Anyway, guns have a use-by date, when ammunition manufacturing and machining come to an end, so the future regarding weapons will be medieval. A quarter staff, in the future will be high tech.

Added to this is the theme of survivalism, detailed in my book, *The Barbarian Reborn*, which I hope everybody has bought in multiple copies, being one of the best books on this topic. Anyway, we are all survivalists now, and the mainstream media which once opened its cloaca and sprayed diarrhea over the idea, has come on board thanks to a mysterious unnamed virus that corporations will not allow discussion of.[43]

The final section may be a surprising addition, advancing a case for mathematical skepticism, epistemological and logical arguments about the insecurity of the foundations of mathematics, as an attack on universalism and rationalism. On this I received considerable help from a pretty red head with a PhD who did not want to lose her job by doing it herself, but still enjoyed a good bang. If she was in physics, I would have said, a "Big Bang." Anyway, mathematics is the poster child of science, of those who worship high IQs and technological sophistication, a symbol of civilization, and yes, I am thinking of you Jared. Naturally, a barbarian such as myself would be eager to engage in a refutation of their central claims, and knowing next to nothing about anything should not halt one. Frankly, I am tired of all of this pompous symbolic bullshit, and just for the intellectual barbarian and iconoclastic fun of it, lash out with semi-informed polemics at the idea that there is some Platonic realm of number and that reality is ultimately mathematical,[44] and even that the entire system is consistent. These essays gradually evolved into a sustained skeptical cranky critique of the foundations of

---

[43]   https://newrepublic.com/article/157635/were-preppers-now-notes-apocalypse-mark-o-connell-book-review.

[44]   Max Tegmark, *Our Mathematical Universe: My Quest for the Ultimate Nature of Reality*, (Borzoi/Alfred A. Knopf, New York, 2014).

mathematics, for the sheer intellectual iconoclastic barbarian hell of it, and to say "fuck" and other polite words in a mathematical context. Sure, mathematics is useful in technology as a fictional anti-realist fictionist "philosophy as if" (Hans Vaihinger), especially in creating weapons to destroy one's enemies, that I grant, but I deny its privileged place in the pantheon of knowledge, as an exemplar of universalism and rationalistic "truth," because there is no such thing; I agree with Nietzsche on that one; perspectivism everywhere. It is as much a myth and fiction as everything else. If black magic "worked," as the ultimate pragmatist and relativist, I would try that too. Truth is a poor second to the will to power. And, am I "right"? Well, I don't care a fuck either way, it is way too far down the tracks now.

And I should add that while relativism and postmodernist deconstructions are all the go, relativism was only a problem because the West wanted to assert itself as a rationally justified universal culture.

As stated, this is not a systematic treatise, but a collection of essays in preparation is a more systematic work dealing with the Odinic survivalist philosophy in detail. Here, it merely wanders in and out of the narrative, much as Odin wanders. Or, maybe I will just get bored with the whole writing game, press "delete" and train in silence, eagerly waiting for Ragnarök. I guess sales will determine the future. Over to you guys, and your generous credit/debit cards.

PART I

AGAINST CHRISTIANITY

I. CHRISTINSANITY

Cracking the Christian Nut

Christianity has emptied Valhalla, felled the sacred groves, extirpated the national imagery as a shameful superstition, as a devilish poison, and given us instead the imagery of a nation whose climate, laws, culture, and interests are strange to us and whose history has no connection whatever with our own.

- G. W. F. Hegel *The Positivity of the Christian Religion*, (1795)

INTRODUCTION

Mainstream Christianity has fallen on hard times, with declining numbers of people professing to be Christian across the West.[45] At the same time, the mainstream church is continuously rocked by kiddy fiddling offenses, some of which go right to the top of the rotten tree.[46]

The Christian fundamentalists, defined as those believing that the Bible is true in all statements, or in all substantial doctrinal statements, say that all this shows is that the majority of people are not "Christians," like them. They say that they are living "just like Jesus," and if the West had not strayed from the literal word, then the

---

[45] http://www.telegraph.co.uk/news/2017/09/04/britain-has-non-believers-ever-church-england-christians-make/.

[46] "Pope Silent on Claim He Ignored Abuse," August 27, 2018, at https://www.breitbart.com/news/pope-silent-on-claim-he-ignored-abuse/.

West would not face the decay it now faces, as God's punishment. If Christianity dies in the West, then the West will die. Yet, that is precisely what is happening, as the next "Christianity" looks like being a Third World globalist religion, that does not share much with the mother's milk religion our Southern friends grew up with at Sunday school.[47]

The demographic displacement, accepted by a universalist church, will seal the doom of Christianity, barring divine intervention, which is always the fundamentalist's *deus ex machina*.[48]

The work of "Satan"? Maybe, but whatever, it is coming when the West falls.

WN TO THE RESCUE?

The discussions of Christianity in WN ('white nationalist') publications, are usually very restrained and controlled. For example, MacDonald in "Christianity and the Ethnic Suicide of the West,"[49] says that Christianity was the dominant religion during Western European expansion, and "the decline of the West has co-occurred with the decline of religion among Western elites." MacDonald is even more explicit in his review of, and preface to, Giles Corey, *The Sword of Christ* (2020),[50] where he says, "Until quite recently, the flourishing of science, technology, and art occurred entirely within a Christian context." And: "Christianity *per se* is certainly not the problem."

---

[47]   https://www.americamagazine.org/issue/420/article/next-christianity; https://www.amazon.com/Next-Christendom-Coming-Global-Christianity/dp/019518307X.

[48]   http://www.spiegel.de/international/spiegel/christianity-in-africa-jesus-in-the-morning-voodoo-in-the-evening-a-463787.html; http://www.neworleansonline.com/neworleans/multicultural/multiculturaltraditions/voodoo.html;http://room5.trivago.com/4-places-voodoo-queens-rule/.

[49]   http://www.theoccidentalobserver.net/2015/04/27/christianity-and-the-ethnic-suicide-of-the-west/.

[50]   Kevin MacDonald, "Preface to Giles Corey's The Sword of Christ," https://www.counter-currents.com/2020/08/kevin-macdonalds-preface-to-giles-coreys-the-sword-of-christ/.

The fallacy of statistical inference committed here is that correlation does not prove causality; the two phenomena could have a third factor causing them, especially since other civilizations, such as ancient Greece, Rome and China, arose on a non-Christian foundation. This is a variant of "the West, by definition is Christian" argument, and it is fallacious. There is no doubt that Christianity strongly influenced the development of Western civilization, but so did paganism: Plato, Aristotle, Homer, and in the modern era, most past philosophers. The fundamentalists could condemn all of this as "Satanism," but all they are really doing is redefining their terms to question beggingly suit them. Logic, a Greek invention, was never a Christian specialty.[51] There is a counter-argument to MacDonald's position:

> And yet, we should not forget that the Western world did not begin with the birth of Christ. Neither did the religions of ancient Europeans see the first light of the day with Moses—in the desert. Nor did our much-vaunted democracy begin with the period of Enlightenment or with the proclamation of American independence. Democracy and independence—all of this existed in ancient Greece, albeit in its own unique social and religious context. Our Greco-Roman ancestors, our predecessors who roamed the woods of central and northern Europe, also believed in honor, justice, and virtue, although they attached to these notions a radically different meaning. Attempting to judge, therefore, ancient European political and religious manifestations through the lens of our ethnocentric and reductionist glasses could mean losing sight of how much we have departed from our ancient heritage, as well as forgetting that modern intellectual epistemology and methodology have been greatly influenced by the Bible.

Just because we profess historical optimism — or believe in the progress of the modem "therapeutic state" — does not necessarily mean that our society is indeed the "best of all worlds." Who knows, with the death of communism, with the exhaustion of liberalism, with the visible depletion of the congregations in churches and synagogues, we may be witnessing the dawn of neopaganism, a new blossoming of old cultures, a return to the roots that are directly tied to our ancient European precursors.

...

Great lamenting is heard from all quarters of our disenchanted and barren world today. Gods seem to have departed, as Nietzsche predicted a century ago, ideologies are dead, and liberalism hardly seems capable of providing man with enduring spiritual support. Maybe the time has come to search for other paradigms? Perhaps the moment is ripe, as Alain de Benoist would argue, to envision another cultural and spiritual revolution—a revolution that might well embody our pre-Christian European pagan heritage?[52]

Interestingly enough, many of the founding Fathers of America, were deists, not strictly Christians at all. Thomas Jefferson, the third president of the United States, was probably a deist too, and in his letter to Thomas Cooper, February 10, 1814, said that the US common law came from England, originating in the middle of the fifth century, but Christianity was not introduced until the seventh century, so the common law did not have a Christian origin, only influence: "the common law existed while the Anglo-Saxons were yet pagans, at a time when they had never heard the name of Christ pronounced or knew that such a character existed." Jefferson wanted the great seal of the United States to bear on one side the images of Hengist and Horsa, the pagan Saxon leaders who settled England.

Further, the controversial thesis of Edward Gibbon, *The Decline and Fall of the Roman Empire* (1776), is that Christianity, along with the invasion and settlement of barbarians of Germany and Scythia, as well as other factors, led to the decline and fall of Rome. If Christianity once took down Western civilization (with help), it can do so again. It must be doing something wrong!

---

52   As above.

Along with all of this, Christianity itself was modified by its interaction with paganism detailed in James C. Russell's *The Germanization of Early Medieval Christianity.*

MacDonald rightly notes that the non-Christian Left and others today are the real champions of universalism and anti-Westernism, which is true. It does not follow from that that Christianity has not had its historical part to play, especially given that Christianity provided the metaphysical foundations for liberalism, a liberalism that was arguably flawed from the beginning, with the worms of decay already in its heart.

MacDonald also argues that the frequent claim made by WNs that Christianity is an alien import because of its universalism, and proneness to guilt, cannot be correct, because these elements are not seen in other Middle Eastern religions. True, but again a *non sequitur*: these destructive elements were absent in native Northern European religions, and only arose with Christianity. Nothing here shows that Christianity is not riddled with such elements; indeed, even if this was true in the distant past (strong men like the Conquistadors hardly "lived like Christ"), it is not true today, as Christianity, and every other decadent force drowns the West in a sea of promiscuous, pathological altruism.[53] These problems with Christianity are bravely admitted by the likes of sensible defenders such as Giles Corey in *The Sword of Christ* (2020), who wants to see a return to premodern Christianity, with a complete rejection of liberalism and all its follies. But, the worms will still lie in the heart of things.

We also find a soft response to the Christian question at other WN sites,[54] where Christianity is supposed to be at odds with the legalism of Judaism, and hence is radically different. Irrelevant; if doctrines are not identical there will always be points of difference, and while it is indeed true that Christianity lacks the legalism of Judaism, that alone is not reason for accepting Christianity.

---

[53]   http://www.theoccidentalobserver.net/2015/04/28/drowning-in-altruism
-thoughts-on-white-pathology-and-the-invasion-of-europe/.

[54]   https://www.counter-currents.com/2018/01/a-mythology-for-the-new-right/.

In another well-quoted piece,[55] the author argues that Christianity is not the main force driving the Western European cultural suicide today, and political strategy depends upon not upsetting Christians and working together. Here in America, I have been engaged in grass roots political opposition since the late 1970s, and my late Texan father began in the post-World War II era, and was active supporting America's pro-Vietnam War effort, himself being a WW II marine. It has been a long frustrating road of defeats because of the stupidity and apathy of the sheeple, but also because of the poor quality of the resistance. We found that the Christians "just don't get it," and are simply beyond reason and argument, on too many things. In my experience, mostly fundamentalist American Christians are generally hard to work with because of their arrogance and the belief that they are "backed by God," and hence are little Gods, and consider people like me as damned. Others are just living confirmation of the Dunning-Kruger effect, of being too dumb to recognize their level of dumbness.

Coady Colinson, in "Can Paganism Save Christian Europeans,"[56] takes an odd compromise path, asserting that Europeans have never stopped being pagans, as seen in the hero tradition in movies and literature, and at present unworldly Christianity is racially dysfunctional, so for survival, it is necessary for a time to led the old Gods rise once more, but this would not be a permanent situation. Really? If Christianity has led Europeans to the brink of extinction, why should we face the prospects of yet another cycle of existential threats? Time to end it.

To finish on the WN politics issue, I think the best refutation of Christianity in politics is unintentionally given by Andrew Fraser in *Dissident Dispatches* (Arktos, London, 2017), which documents Fraser's misguided, but well-meaning attempt to find a renewal of Christianity through study at a theological college. Instead, he finds politically correct liberalism. I felt sorry for the old guy.

---

[55] https://www.counter-currents.com/2017/12the-christian-question-in-white-nationalism-2/.

[56] https://www.eurocanadian.ca/2020/07/can-paganism-save-christian-europeans.html.

But, he should have read William Gayley Simpson, *Which Way Western Man?* (1978), which traced the author's evolution from a fundamentalist Christian to Nietzschean thought: "Christianity unfits any people for very survival. The malady of which the whole White man's world is dying is – Christianity." (p. 83) An exaggeration perhaps, but part of the truth.[57]

CHRISTIANITY AND TRUTH

All of the above considerations are beside the point if Christianity is true in a common-sense realist way of correspondence to reality, being that which exists independent of mind and society and is thus not a social construction. We are thus thinking about Christianity being true in the sense that its core doctrinal statements of morality and historical fact are true. In other words, Christianity should be capable of a philosophical defense of its core doctrines, and it should have a sound core of historical truth. Those of my critics who say that all non-Biblical barbarians like me are cucked because we don't have the strength to "live like Jesus," would have their position severely damaged if it was the case that Jesus had/did not exist (mythicism), or was not "God."

Philosophy first. The theological debates about arguments for the existence and non-existence of God continue with no resolution, which is not surprising because nothing in philosophy is ever really resolved, making this truly an intellectual masturbation discipline. As far as it is possible to approach things from a neutral position, a free thinker probably should be agnostic about the existence of the philosopher's God because of the unending debate alone. The most challenging arguments against God's existence involve logical incompatibilities in the divine attributes,[58] but there is no point going into such technicalities here. Who will care about Cantor's power set theorem, and other technicalities? Wait until later in this book.

---

[57] https://www.counter-currents.com/2013/07/why-christianity-cant-save-us/.

[58] https://www.amazon.com/Incomplete-Universe-Totality-Knowledge-Truth/dp/0262519119.

Perhaps the more interesting philosophical arguments relate to the problem of evil is why an all-powerful God allows evil to exist in the world. There have been forests of trees felled to publish dissertations about this question. The philosophers generally approach this in the abstract and apologists argue that free will justifies the existence of evil for various soul-strengthening reasons. We could grant this, but turn to the harder problem of evil, why so much evil is allowed even given all of these excuses? Would sacred freedom be violated if one less misery did not occur in the world? This is the problem of gratuitous evil.[59]

The fanatical Christian apologist usually responses to this by invoking the hypothesis of Satan, a fallen angel, less powerful than God, who seems to play havoc in the world, exercising his free will. Lucifer fell because he rebelled against God, probably a domestic dispute over the creation of inferior beings, humans. Some argue that natural evil that pre-existed Adam (death and destruction, the second law of thermodynamics), thus came into being. Yet, the postulation of Satan is utterly *ad hoc*, done only to save the system from falsification. But, allow it for the sake of argument. There is then a new problem, like the "why is there something rather than nothing problem," facing secular cosmologies: why would an all-knowing God have created the entire system at all, knowing that free will would allow evil to enter reality? Why not remain in communion with the rest of the contradictory elements of the trinity; logically distinct, but different?

The humanists, and the WNs, frequently raise the internal problem of evil, that the Old Testament is full of genocide and the destruction of enemies. I, for one, do not have the same "shock, horror," as some do. And, I do not have the hatred for the Old Testament that some atheists/closet anti-Semites seem to have, for like Nietzsche I see this grand book as far superior as a warrior theology to the anaemic, puny, New Testament. The Old Testament, as literature rises to heights of excellence that the New Testament never reaches, and certainly deserves respect, in my opinion. The

---

59   https://infidels.org/library/modern/nontheism/atheism/evil.html.

stories of the survivalist Noah (Genesis 5-9) and Samson (Judges 13-16), are truly epic tales. In general, I agree with Nietzsche, that Judaism is a far superior religion to Christianity, in any of its forms. But, respect aside, it is not my religion.

However, literature, the Old Testament is, which taken in context is not necessarily bad. The numerous Biblical passages asserting the destruction of enemies are, sadly, almost all false, a creation of "nation building" ideology among the ancient Hebrews, which I don't see any problem with, and hope that my tribe can learn from this stark realism in the post-apocalyptic wastelands to come. The vast literature condemning this exercise in group survival, written from some illusionary Christian and liberal high moral ground, is therefore misplaced in my opinion.

Nevertheless, even if we pass over the moral issue, there is still a logical and epistemological problem arising from the archaeological critique of the Old Testament, namely that its key historical claims and figures, all believed by Jesus, are not true, and did not exist, as argued by I. Finkelstein and N. A. Silberman, *The Bible Unearthed*.[60] No Israelites in Egypt, no wandering in the desert, no military conquests, no united monarchy of David and Solomon.[61]

A growing volume of evidence concerning Egyptian border defenses, desert sites where the fleeing Israelites supposedly camped, etc., indicates that the flight from Egypt did not occur in the thirteenth century before Christ; it never occurred at all. Although Johnson writes that the story of Moses had to be true because it "was beyond the power of the human mind to invent," we now know that Moses was no more historically real than Abraham before him. Although Johnson adds that Joshua, Mose's lieutenant, "began and to a great extent completed the

[60] I. Finkelstein and N. A. Silberman, *The Bible Unearthed: Archaeology's New Vision of Ancient Israel and the Origin of its Sacred Texts*, (Free Press, New York, 2001).

[61] Ze'ev Herzog, "Deconstructing the Walls of Jericho," *Ha'aretz Magazine*, October 29, 1999, pp. 6-8: http://www.umich.edu/~proflame/neh/arch.htm; M. Sturgis, *It Ain't Necessarily So: Investigating the Truth of the Biblical Past*, (Headline, London, 2001).

conquest of Canaan," the Old Testament account of that conquest turns out to be fictional as well. And although Johnson goes on to inform his readers that after bottling up the Philistines in a narrow coastal strip, King David "then moved east, south and north, establishing his authority over Ammon, Moab, Edom, Aram-Zobar and even Aram-Damascus in the far north-east," archaeologists believe that David was not a mighty potentate whose power was felt from the Nile to the Euphrates, but rather a freebooter who carved out what was at most a small duchy in the southern highlands around Jerusalem and Hebron. Indeed, the chief disagreement among scholars nowadays is between those who hold that David was a petty hilltop chieftain whose writ extended no more than a few miles in any direction and a small but vociferous band of "biblical minimalists" who maintain that he never existed at all.

In classic Copernican fashion, a new generation of archaeologists has taken everything its teachers said about ancient Israel and stood it on its head. Two myths are being dismantled as a consequence: one concerning the origins of ancient Israel and the other concerning the relationship between the Bible and science.[62]

The lack of proof of the existence of say, King David, has devastating consequences for our Bible-believers, because key things which Jesus (= God) believed are false, which is logically inconsistent with the assumption of his divinity and indeed identity with God. There is simply no way around this if the standard rules of evidence are held to. A divine being cannot believe falsehoods.

There is also the parallel problem of the multitude of contradictions in the Bible, and apologists usually sharpen their skills at explaining each of these away. These contradictions have been discussed for well over a thousand years, but a particularly

---

[62] D. Lazare, "False Testament: Archaeology Refutes the Bible's Claim to History – Criticism," *Harper's Magazine*, March 2002, at: http://www.yorku.ca/dcarveth/false_testament.

clear American response to them is given by Thomas Paine in *The Age of Reason*, (Paris, 1794), and R. W. Hinton, *Arsenal for Skeptics*, (A. S. Barnes, New York, 1961), is very good, with numerous references to literature not much reviewed today. One example is the contradictory accounts of Jesus's birth in the Gospels of Matthew and Luke.[63]

The answer given by theologians since the Enlightenment is that the Bible is a historical construct, complied from sources, and not dictated or handed down by God, so, contradictions show this.[64]

The Gospels being human constructs, written 40 to 60 years after the death of Jesus, if he existed at all, are not eyewitness accounts (how would we know this today anyway?), and hermeneutic evidence indicates that the Gospels were not written by the alleged traditional authors.[65] Thus, Matthew, alleged to be an eyewitness account of the life of Jesus, copies 80 percent of the text of Mark, but the traditional author of Mark was not an eyewitness or a disciple, which is inconsistent for an alleged eyewitness account.[66] They would go down for plagiarism if this was a student term report.

No one who was a disciple would have written a Gospel, "According to..." as an eye witness would have deposed the contents in the first-person narrative.

Finally, there is some downright embarrassing stuff for orthodox Christians which does not particularly bother me; there is evidence that Jehovah had a female consort called Asherah, a fertility goddess.[67] As an Odinist, I see nothing wrong with that, and wish the happy couple well, but orthodox Christians may beg to differ. Inscriptions

---

[63] https://infidels.org/library/modern/matthew_ferguson/gospel-genre.html.

[64] Michael Martin, *Atheism: A Philosophical Justification*, (Temple University Press, Philadelphia, 1992); Richard Carrier, *On the Historicity of Jesus: Why We Might Have Reason for Doubt*, (Sheffield Phoenix Press, Sheffield, 2014).

[65] https://celsus.blog/2013/12/17/why-scholars-doubt-the-traditional-authors-of-the-gospels/; http://www.rationalrevolution.net/articles/gospel_mark.htm.

[66] https://www.counter-currents.com/2017/07/is-the-gospel-according-to-mark-an-allegory/; http://jesusneverexisted.com.

[67] J. Rose, *The Myths of Zionism*, (Pluto Press, London, 2004), p. 23.

found in ancient Hebrew in the late 1960s in excavations at Khirbet, el-Kom, near Hebron, say: "Blessed...by Yahweh...and his Asherah." A second site found a similar expression in the late 1970s at Kintillet Ajud in north-eastern Sinai. This indicates that there was a socio-historical process occurring producing the Bible as we know it.

CONCLUSION

Taking everything here, the balance of reason is against the truth of fundamentalist Christianity. There are no good reasons for accepting the Bible as true. The sheer volume of criticism counts against this. But, this is all water off a duck's back for the American Christian fundamentalist who by definition never accepts any evidence of any problem as showing a fault in their world view. Just believe harder, or Jesus works in mysterious ways. We have heard this sophistry all before. The replies were written before even reading the references, because they just MUST be wrong, because sweet me was brought up by mom to believe.

I hope that Christians and liberals are consistent at a personal level with their "brotherhood of man" and egalitarianism when the collapse comes. But, given the hypocrisy of both liberals and Christians, I imagine they will drop all the ideology and run for the hills. We will see how strong they are striving to live like Jesus and die on their own crosses. I do not blame them, but remember: by rights, they should be giving and living like Jesus until they perish.

II. AGAINST CHRISTIAN WN, ANTI-GERMANIC PAGAN BIGOTRY

Anumber of white nationalist and manosphere sites have been attacking primarily Germanic paganism. The attacks, based on little or no scholarship, maintain that Nordic paganism in general, and specifically Germanic paganism, in prehistory, supported feminism and homosexual practices. Only a reformed Christianity can save the West. In this chapter I respond to the white nationalist attack upon my tribe. Of course, nothing ill is implied against homosexuality today – this is all historical. As usual, Jack Donovan, himself a homosexual pagan, has an insightful take on the homosexuality issue: *Androphilia: Rejecting the Gay Identity, Reclaiming Masculinity* (Dissonant Hum, 2012), which seems to me to be the correct position for pagans to take. Homophobia is very much a by-product of the Christian *weltanschauung*, and once the Christian position is dropped, along with it goes many of its hang-ups and conceptual baggage. In a world on the brink of destruction, I simply do not regard the homosexuality issue as one of concern. Nevertheless, for the sake of critique of the white nationalist position, we will go with their assumptions, rather than end matters quickly the Jack Donovan way.

As stated, the argument immediately reveals itself as profoundly illogical, since it could, and is argued, that even if there were these defects in paganism in pre-history, pre-history is past, and nothing prevents a new neo-paganism from not making the same mistakes, as Alain de Benoist argues in *On Being a Pagan*.[68]

---

[68] https://www.amazon.com/Being-Pagan-Alain-Benoist/dp/0972029222.

Here is a brief description of this book that anticipates the recent Christian white nationalist critique, and slams it:

> In this small masterpiece, the great French thinker Alain de Benoist claims that only the pagan deities of ancient Europe offer a spiritual recourse to the present religious malaise. The guilt, the fear, the narrow petty-bourgeois obsession with well-being, and the self-loathing love of the Other that has left Western man defenseless before the destructive behaviors of our nihilist age derive from the alien belief system that Christianity introduced to the West. They are not part of the pagan spirit that lives still in the *Rig Veda*, the *Iliad*, or the *Eddas*. Benoist helps us rediscover these ancient wellsprings and the fonts from which future greatnesses may again flow. But let the reader be warned, his *On Being a Pagan* proposes no folkloric or New Age "return to the past," but rather a Nietzschean recurrence in which the future bears all the promise of our distant origins—and thus of another great beginning.[69]

Christian white nationalism is a mistaken position which deserves critique. Paganism may have failed in the past as a survivalist religion, but so has Christianity today, which is imploding. If there is no hope of going back to pre-Christian belief systems, it is equally as hopeless to put one's faith in Christianity to save us. Here, I will be primarily criticizing the anti-pagan arguments rather than attacking Christianity first hand.

The critics, are no doubt aware that there is no institution more corrupted at present with political correctness than the Christian church, with Protestantism being a form of social welfarism, and Catholicism, under the no-hope Pope, championing open borders immigration, with everyone having a right to immigrate to better their lives.[70] Hello the tragedy of the commons, and the quick fall of the West.[71]

---

[69] Michael O'Meara, author of *New Culture, New Right* (First Books, 2004).

[70] https://thenewright.news/2017/07/robert-spencer-pope-francis-defender-of-islam/.

[71] http://science.sciencemag.org/content/sci/162/3859/1243.full.pdf.

As well, we need not dwell on the fact that homosexuality exists right throughout the church, and it goes right to the top, with F. Martel's book, *In the Closet of the Vatican*,[72] claiming that 80 percent of Vatican clergy are homosexuals, which is not to say that this is bad, of course. Christian white nationalists say that this is all modern, although they are not willing to cut any slack to Nordic pagans in this regard, not that we need their slack, or cuts.

Nevertheless, homosexual practices have long been part of monastic practice, despite being officially condemned.[73]

THE ARGUMENT FROM ANCIENT GREECE AND ROME

Christian white nationalist opponents sometimes have a sympathetic attitude to ancient Greek and Roman paganism, even while Greece and Roman at their peaks had a substantial degree of Nordic blood, and even though female goddesses were part of their theology, parallel to the Nordic religions.[74]

It is also worth noting that one scholarly book, Peter Heather, *The Fall of the Roman Empire: A New History of Rome and the Barbarians*,[75] presents the argument that Rome did not collapse from the sorts of things that conservatives whinge about, such as high taxes, but was brought down solely by the barbarians. As well, the historian Edward Gibbon (1737-1794), in his *The History of the Decline and Fall of the Roman Empire*, famously proposed that Christianity had a major impact in undermining Rome.[76] This is

---

[72] F. Martel, *In the Closet of the Vatican: Power, Homosexuality, Hypocrisy*, (Bloomsbury Continuum, London, 2019).

[73] https://en.wikipedia.org/wiki/History_of_Christianity_and_homosexuality; http://www.huffingtonpost.com/2013/07/31/ancient-christian-church-gay-marriages_n_3678315.html;https://www.amazon.com/Same-Sex-Unions-Premodern-Europe-Boswell/dp/0679751645; https://www.jstor.org/stable/2865807?seq=1#page_scan_tab_contents.

[74] http://www.dummies.com/education/history/world-history/gods-and-goddesses-of-greek-and-roman-mythology/.

[75] https://www.amazon.com/Fall-Roman-Empire-History-Barbarians/dp/1522670149.

[76] https://www.ccel.org/g/gibbon/decline/volume1/chap15.htm.

quite consistent with Heather's work, as Christianity was the internal force weakening Rome and the barbarian tribes, the external force, ultimately destroying it.

Although there were at times official condemnations, their new class elites freely engaged in homosexual practices, with leading philosophers such as Socrates, losing his supposed reason in the company of "beautiful boys."[77] Whatever; we pagans care not.

It was not just the elites who practiced homosexuality; graffiti at the ill-fated city of Pompeii was full of homosexual comments:

Weep, you girls. My penis has given you up. Now it penetrates men's behinds. Goodbye, wondrous femininity.

It is not clear why this lamenting soul did not pursue bisexuality; perhaps it was a trolling comment. But, now he is dust, cock too.

Similar comments can be found in ancient Roman sources, many also illustrating sexual confusion.[78] This is not to say that homosexuality was accepted by the ancient Greeks, say, as there is evidence presented by Adonis Georgiades, *Homosexuality in Ancient Greece: The Myth is Collapsing*, (2004), that the evidence that homosexuality was accepted by the ancient Greeks and Romans, is scanty. The point to be made is that there is no more evidence of this than there is for such acceptance in ancient Germanic culture, for no doubt there can be cherry picking of examples, which does not prove a general case.

So, here is the first problem for the conservative Christian white nationalist. Given their anti-homosexualism, they will have to drop ancient Greece and Rome form their hit parade because the same argument which they use against Nordic paganism, undermines their championing of the civilizational virtues of ancient Greece and Rome, which is a *reductio ad absurdum* of their argument.

---

[77]   http://www.livius.org/articles/concept/greekhomosexuality/; http://www.faculty.umb.edu/gary_zabel/Courses/Morals%20and%20Law/M+L/Plato/homosex.htm.

[78]   https://en.wikipedia.org/wiki/Homosexuality_in_ancient_Rome.

IRON AGE NORDIC WARRIORS

Let us step back a little further in time. A post "Hyper-Masculine Behavior Among Iron Age Scandinavian Men,"[79] makes some useful quotations from the scholarly work, by Lotte Hedeager, *Iron Age Myth and Materiality: An Archeology of Scandinavia, AD 400–1000*, (London: Routledge, 2011), which discusses hyper-masculinity among north-western Europeans before Christianity. As the article says:

> It paints a picture of a hyper-masculine, completely militarized society in which male sexual penetration was a marker of power, while being penetrated was, for a male, the ultimate insult. Accusing a man of having been sodomized was a grievous accusation, with the same penalty as for murder. Older males lacking the power or ability to penetrate took on the status of women and were even ridiculed by slaves. Women were spoils of warfare and raiding.

Here is what the book says about the power of penetration:

> In the Norwegian Gulathings law, outlawry was the penalty if a man accused another of being *sannsorðenn* (provably sodomised). Also, full personal compensation must be paid if a person says to another man that he has given birth to a child. The third is if he compares him to a mare, or calls him a bitch or a harlot, or compares him with the female of any kind of animal … Then he can also kill the man as an outlaw as a payback for those words that I have now spoken, if he takes a witness to them.

> It is commonly accepted among scholars that *níð* was not a question of biological reality (after all, pregnant and childbearing men are metaphorical constructions); it was instead a sophisticated form of gendered insult, to be equated

---

[79] http://www.theoccidentalobserver.net/2016/01/14/hyper-masculine-behavior-among-iron-age-scandinavian-men/.

with the "murder" of someone's honour. That is why *nið* has the secondary meaning of death. The conceptualization of *nið* is aimed at the person who was suspected of being the object of sexual penetration, whether man, woman, or animal. The masculinity of the practitioner is not the moral problem. In Old Norse society the physical act of penetration had no moral connotations, neither if one man penetrates another, turning his anus into a vagina (and metaphorically making him pregnant), nor if he practiced sodomy, called *tidelag*. What was deeply defamatory, however, was to accuse a man of having being subject to penetration by another man – or a male animal – or of being transformed into a female or a female animal. The *nið* was subjected to a transformation into "female," not specifically into an animal. In short, *nið* is an accusation of unmanliness and softness, that is, the person is *argr* (*ergi, ergjask, ragr*, etc.) sexual terminology as a mark of identity, although the word may relate to a practice within a fluid sexual system.

Thus, the ancient Germanics were hardly the sex/gender modernists that the contemporary Christian white nationalist critics say they were; their position was quite complicated. Whether this is good or bad is irrelevant today, but it refutes their anti-pagan Christianity.

THE QUESTION OF THE CELTS AND GERMANICS

The strongest arguments for the position that the ancient barbarians embraced feminism and homosexualism, comes from a consideration of the Celts. It is a mistake though to equate their culture and religions such as druidism, with that of the Germanic peoples.[80]

One of the main sources quoted against the Celts is by Diodorus of Sicily (400 BCE), who said:

---

[80] https://www.amazon.com/Germanic-People-Origin-Expansion-Culture/dp/0880295791; https://www.amazon.com/Germanic-People-Origin-Expansion-Culture/dp/0880295791.

they [Celts] have beautiful women but pay no attention to them – rather they weave around other males in a strange frenzy. They are accustomed to sleeping on the ground upon hides of wild beasts and wallow together with male partners on both sides for fucking. And most paradoxically ... they do not regard this as a disgrace; rather the opposite – whenever their freely-offered gift of sexual gratification is not received favorably, they regard it as a dishonor.

The frequently quoted passage is similar to the type of comments made by Hippocrates (460-370 BCE), against the Scythians, who were also viewed as hostile barbarians and likewise vilified, as a rhetorical strategy.[81] It was probably just an early form of hate speech.

That being so, scholarship does not support the position that the Celts had a matriarchy, and most contemporary "feminist scholars," whatever that means, accept that although women had more rights than in ancient Greece and Rome, Celtic society was "male-dominated."[82] Celtic society was patriarchal, with care for children the primary role for women, and Celtic women did not have legal equality with men. That no more makes them feminists, as did the existence of Queens in pre-modern Europe make Christians at the time feminists, although many Christians of that age did object to female monarchs.[83]

What about the Germanic barbarians? Here there is even less of a case for the acceptance of feminism and homosexualism, as documented by Roman historian Tacitus (58-120 AD), in *Germania*. This book proclaimed the warrior virtues of the German barbarians, who "have never contaminated themselves by intermarriage with foreigners but remain of pure blood, distinct and unlike any other nation." He describes how traitors were hung on trees and

---

[81] https://en.wikipedia.org/wiki/Scythians.

[82] https://en.wikipedia.org/wiki/Ancient_Celtic_women;http://www.celtlearn.org/pdfs/women.pdf.

[83] https://en.wikipedia.org/wiki/The_First_Blast_of_the_Trumpet_Against_the_Monstruous_Regiment_of_Women.

homosexuals drowned in bogs.[84] But, we need to be careful with this book since Tacitus was wanting to critique his society and may have got into romanticism, or even false news, who knows, but his heart was in the right place.

ZOOPHILIA IN THE ANCIENT WORLD

The white nationalist types who wax lyric about ancient Greece and Rome, while pissing on the Germanics, need to explain things like this:

> The ancient Romans brought bestiality to an art form. Roman women were known to have kept snakes for sexual purposes. Bestiality was even a center of attraction at the Coliseum and the Circus Maximus, where men and women were brought in to be raped by animals.
>
> Unlike the ancient Romans, however, who practiced bestiality for pleasure or entertainment, the ancient Greeks were into bestiality for religious reasons, turning it into a central rite during the Bacchanalia and part of the ceremony at the Temple of Aphrodite Parne. But, according to ancient writers, the Egyptians might have taken the bestial cake. It was said that there was nothing more common in ancient Egypt than young women having intercourse with bucks. According to Herodotus, goats were said to be part of the religious practice in the Temple at Mendes. Crocodiles were not even excused when it came to bestiality—ancient Egyptian crocodile hunters supposedly had sex with female crocodiles before killing them.[85]

Your guess is as good as mine about how they performed such acts. Probably … carefully.

[84] https://vultureofcritique.wordpress.com/2015/06/07/would-the-pagan-germans-drown-oscar-wilde-in-a-bog/.

[85] http://listverse.com/2015/01/16/10-bizarre-sex-facts-from-the-ancient-world/; http://news.cornell.edu/stories/2011/09/annetta-alexandridis-lectures-bestiality.

Pagan Vikings regarded being involuntarily sodomized as an insult and an excuse for a dual: B. Hubbard, *The Viking Warrior: The Norse Raiders Who Terrorised Europe*, (Amber Books, London, 2015). The only piece of "evidence" against them may come from goddesses being present in their theology, but here, all we have to rely upon are the records made by Christians who dearly wanted to finish of the old religion. The evidence of shield maidens is weak, but if they existed, they were insignificant.

But, sometimes a glimmer of the real past creeps through the Christian censor, as in this passage from the *Havamal*, verse 83, the words of Odin our one-eyed God:

The speech of a maiden should no man trust
nor the words which a woman says;
for their hearts were shaped on a whirling wheel
and falsehood fixed in their breasts.

If this is objectionable to feminist sentiments, then no doubt Willie the Shake's lines are too:

Could I find out
The woman's part in me—for there's no motion
That tends to vice in man, but I affirm
It is the woman's part; be it lying, note it,
The woman's; flattering, hers; deceiving, hers;
Lust and rank thoughts, hers; revenges, hers;
Ambitions, covetings, change of prides, disdain,
Nice longing, slanders, mutability,
All faults that name, nay, that hell knows,
Why, hers, in part or all; but rather, all;
For even to vice
They are not constant, but are changing still . . .

- *Cymbeline* Act 2, scene 5, 19–30.

III. CRUCIFIED CHRISTIANITY

There are some "good" articles online showing how the mainstream Christian mind works. But, Christianity is a vector that carries a pathogenic mind virus, that first spawned liberalism, and now globalism and cosmopolitanism, grounded in open borders capitalism. Once one drops the assumption that Christianity is true, then there is the freedom to see that Western civilization arose despite it, like a kite against the wind, and that this belief system has merely clung to the European West.

There are many good popular books that could help in deprogramming, and we will cover them, in diatribes against the cult, to come.

It is not necessary to do a theology degree to uncover the near infinite difficulties present in the scriptures. The short of the long of a modern theology degree though, is that the Bible is socio-historical construction, written by numerous people over long periods of time, drawing on literary material and myths of the age, and packaged for mass consumption by largely scientifically ignorant people. The Old Testament has little ancient historical accuracy and is mainly an attempt to forge an identity for the Jewish people.[86]

There are insuperable problems with the Christian bag of tricks, right from the creation mythology, through to the scriptures and their historicity and internal coherence. The bottom line on the creation issue is that even if we grant that evolution has technical limitations, no problem facing evolution (e.g. say the improbability of biological complexity arising from natural selection of random

---

[86]   I. Finkelstein and N. Silberman, *The Bible Unearthed*, (Free Press, 2002).

genetic mutations), makes Christian creation plausible. There is an infinity of other hypotheses, also *ad hoc*, which could account for the existence of the world, such as its emergence from another universe (multiverse hypothesis), or even the creation by a god who is beyond good and evil. In fact, such a god does not face the problem of evil which the Christian God faces.

Usually the problem of evil, in philosophy classes, is to attempt to reconcile the existence of an all-good, all-powerful, all-knowing God, with the existence of evil, either moral evil, or natural evil. Moral evil is accounted for by the freewill defense, embodied in a flawed way in the garden of Eden story. The problem with the Garden of Eden story is that before eating of the fruits of knowledge of good and evil, Adam and Eve were morally innocent, like young children, so it seems immoral of God to punish them. Christians will pass over this uncomfortable issue, saying that it was still disobedience, although neither of the naked pair knew that not following orders was wrong. This is like punishing a two-year old for an offence against the tax code.

The Old Testament is full of many things completely inconsistent with contemporary moral values, all allegedly supported by God. This is a real problem for liberal Christians.[87] There are literally thousands of Old Testament passages that are completely contrary to modern morality, and so much the worse for modern morality. And, this is the so-called word of God. That is the real problem of evil, and from a rational perspective it cannot be solved without a major challenge to Christian fundamentalism. Christians usually ignore the issue, and atheists have not beaten them hard enough on this issue.

Then, there is the problem of evil in the world before man, which is accounted for by the hypothesis of the fallen angel, Lucifer. Again, having free will, Lucifer was able to disobey, and before you know it, became infinitely evil, *the Satan*. I have heard Christian creationists argue that the second law of thermodynamics arose from Satan's rebellion, as did death e.g. the design of carnivores to eat meat.

---

[87] John Spong, *The Sins of Scripture*, (2006).

Without these acts of rebellion, the tiger's claws would be used to lovingly caress all and sundry. Or, when the Fall occurred, creatures suddenly became carnivores. Take your pick of which story to spin.

The fact of the matter is that from a rational position, all of this is just arbitrary. The more parsimonious hypothesis is that the creator, who may have created *ex nihilo*, or who could even be a F-grade alien scientist, is not all-good, or all-powerful, or all-knowing. That accounts for the universe better than Judeo-Christianity.

In fact, the human body, with its numerous design defects e.g. the design of the male prostate), is hardly a creation reflecting anything divine. It smacks of a slapped-together body, and one where micro-organisms living on it, outnumber the cells of the body, full of design flaws and "scars."

These micro-organisms, we will see, even make a genetic contribution to the human organism, which is not consistent with the idea that the human being is made in the image of god. And, if the slippery Christian then says that the reference is only to the mental and moral aspects, well, that is even worse given the utter madness of most people.

The universe is not consistent with the primitive worldview of Christianity. From the perspective of a theoretical physics, I would rest my case on quantum mechanics alone. Particles which are also waves at the same time, thus having classically inconsistent properties; indeterminacy, and numerous paradoxes from a classical perspective, indicate that we are living either in a cosmic computer simulation (and a poor one at that), or some sort of absurd reality. No rational god would have created the world as modern physics describes things. Hence, there is no Christian God. Existentially, there is a good case that the entire universe is absurd, perhaps in some way illusory and certainly doomed to cosmic oblivion. That is, if it is anything more than a simulation by trans-multiverse bored super-programmers.[88]

---

[88]   https://hackernoon.com/the-great-simulation-why-quantum-physics-artificial-intelligence-and-eastern-mystics-all-agree-b6c185213a18.

BIOLOGICAL PROBLEM FOR CHRISTIANS

Have I got this right? The Christian God made man in his image. Fine; doesn't that mean that humans will therefore be metaphysically special? There will be a division between the human world and the animal world? Humans will have a "soul" manifested in mentation and the creation of cultures. Apparently, that makes us special.

Too bad that whales and dolphins (and other primates) have proto-human-like cultures as well, based on the cognitive capacities produced by the size of their brains:[89]

"The long list of behavioural similarities includes many traits shared with humans and other primates such as:

- complex alliance relationships – working together for mutual benefit;

- social transfer of hunting techniques – teaching how to hunt and using tools;

- cooperative hunting;

- complex vocalizations, including regional group dialects – 'talking' to each other;

- vocal mimicry and 'signature whistles' unique to individuals – using 'name' recognition;

- interspecific cooperation with humans and other species – working with different species;

- alloparenting –looking after youngsters that aren't their own;

- social play."[90]

Basically, human primates came to dominate Earth by means of their capacity to be able to handle weapons, which combined with language/cultural ability, enabled them to gang up on every other fucker who threatened them and kill them off. Higher

[89] https://www.sciencedaily.com/releases/2017/10/171016122201.htm.

[90] https://doi.10.1038/s41559-017-0336-y.

extra-terrestrial life could be more impressed with whales and elephants than the human grunters they might find, and perhaps would delight in exterminating the human rats for sport.[91]

IS GOD A COSMIC MICRO-ORGANISM?

Having previously written about fundamentalist Christianity, I came across this article, which is written from an evolutionary perspective. It represents a challenge to the faith if it was correct: our consciousness, our soul, came from bacteria.

According to two papers published in the journal *Cell*, a virus bound its genetic code to the genome of four-limbed animals. That snippet of code is still very much alive in humans› brains today, where it does the very viral task of packaging up genetic information and sending it from nerve cells to their neighbors in little capsules that look a whole lot like viruses themselves. And these little packages of information might be critical elements of how nerves communicate and reorganize over time — tasks thought to be necessary for higher-order thinking, the researchers said.

Though it may sound surprising that bits of human genetic code come from viruses, it's actually more common than you might think: a review published in *Cell* in 2016 found that between 40 and 80 percent of the human genome arrived from some archaic viral invasion.[92]

As we are all interested in scientific papers, which we read each morning on the crapper, the relevant papers are:

(1) J. Ashly (et al.), "Retrovirus-Like Gag Protein Arc 1 Binds RNA and Traffics Across Synaptic Boutons," *Cell*, vol. 172, 2018, pp. 262-274;

(2) N. F. Parrish and K. Tomonaga, "Endogenized Viral Sequences in Mammals," *Current Opinion in Microbiology*, vol. 31, 2016, pp. 176-183.

---

[91] https://en.wikipedia.org/wiki/The_Predator_(film).

[92] https://www.livescience.com/61627-ancient-virus-brain.htm.

Not only does the human body arise from 40-80 percent of past viral parasitic invasions, incorporating the alien proteins into its being, but the viral proteins have influenced the way consciousness has developed.

Now, if correct, this could be taken as a direct argument for the nihilistic point of view. But, maybe not so quick ... could it be that we were all wrong about thinking of God in humanoid form? Could God be a cosmic virus? Not your garden variety of virus, which would be blasphemous and degrading, but something really grand, something infinite, cosmic and totally uncaring about the slabs of meat it created for the truly important beings—viruses and prions—to feast upon? Maybe the world should be viewed from the top to the bottom, and little things are more important than big things.

The evolutionary biologist J. B. S. Haldane said in his 1949 book *What is Life? The Layman's View of Nature*, p. 248:

> The Creator would appear as endowed with a passion for stars, on the one hand, and for beetles on the other, for the simple reason that there are nearly 300,000 species of beetle known, and perhaps more, as compared with somewhat less than 9,000 species of birds and a little over 10,000 species of mammals. Beetles are actually more numerous than the species of any other insect order. That kind of thing is characteristic of nature.

Haldane got it right about the stars, but wrong about species, for microorganism (bacteria, viruses etc.) rule this world:

> Although the 1998 estimates have been questioned in terms of ocean-dwelling microbes, the University of Georgia researchers suggested that the DRY biomass of bacteria is between 350,000 and 550,000 million tonnes.

> Since the dry biomass of humans is only around 105 million tonnes, the bacteria on Earth weigh at least 3,000 times as much as all of humankind combined.

And I suppose this shouldn't actually surprise us when we stop to reflect on the fact that there are about 50 million bacterial cells in a single gram of soil, and estimates suggest that over 90% of all bacteria on Earth live in the soil.

In fact, someone who may well have had too much time on their hands calculated that the world's soil bacteria weigh as much as the United Kingdom, although working that out couldn't have been easy.

Let's just finish by reminding ourselves how much bacteria you have in—and on—you right now, casting no aspersions on your personal hygiene of course.

The folks at the Human Microbiome Project estimate that all your personal bacteria probably weigh in at between two and six pounds, enough to fill a large soup can, and consisting of something like 100 trillion cells.[93]

Growing bacterial resistance to antibiotics may return medicine to a kind of Dark Age.[94]

The rise of antimicrobial resistance is a global crisis, recognized as one of the greatest threats to health today.

The threat is easy to describe. Antimicrobial resistance is on the rise in every region of the world. We are losing our first-line antibiotics. This makes a broad range of common infections much more difficult to treat.

Second- and third-choice antibiotics are costlier, more toxic, need much longer durations of treatment, and may require administration in intensive care units.

---

[93]   https://ubiomeblog.com/2016/06/27/weighs-bacteria-every-single-person-earth/.

[94]   http://www.who.int/mediacentre/factsheets/fs194/en/; http://www.who.int/dg/speeches/2016/antimicrobial-resistance-un/en/.

Superbugs haunt hospitals and intensive care units all around the world. Gonorrhoea, which is a sexually transmitted disease, is now resistant to multiple classes of drugs. An epidemic of multidrug-resistant typhoid fever is rolling across parts of Asia and Africa.

Even with the best of care, only around 50% of all patients with multi-drug resistant tuberculosis can be cured.

With few replacement products in the R&D pipeline, the world is heading towards a post-antibiotic era in which common infections will once again kill. If current trends continue, sophisticated interventions, like organ transplantation, joint replacements, cancer chemotherapy, and care of pre-term infants, will become more difficult or even too dangerous to undertake.

This may even bring the end of modern medicine as we know it.

As usual, everyone is Fucked.™

CHRISTIANITY VS. BACTERIA

Not to offend all the gentle Christian warriors and bigots out there in US Bible land, but I would like to talk about Christianity and the body. Now, I know, the atheists have given you a hard time on the sex and purity issue, and looking at the disaster their 1960s libertarianism has brought, clearly you have won that intellectual battle, although the shock waves are still eroding society, and alone will probably destroy the West.

What puzzles me though is the biological structure of the world. You tell us from your big brother book, Genesis, that in the beginning, God made the world, and it was good: Genesis 1:13. According to version 1.0 of the creation, nature was created first, then man. Before the Fall, man was also good.

Now what about bacteria? We now know that there are more bacteria on and in our bodies than we have cells. Some are bad, but most play vital roles in human functioning, such as in the gut. Evolutionary theory posits that we are nothing more than a genetic ecosystem made up of the genetic bits and pieces of past organisms. Hence, the general poor "design" of the body, which exhibits the "scars of evolution."[95]

That shit which you proudly forced out this morning, a product of good American beef, fed billions of bacteria, and the brown stuff is mainly bacteria, well up to 54 percent bacterial biomass.[96]

Shit! You didn't know that, did you? Me neither, until I looked it up.

The theological problem is: why would something made in God's image be so "dirty" and "profane," literally crawling in "germs"? This would be so before the Fall too, for bacteria would be doing what they do now. Jesus and Mary, could have had "immaculate" bodies, and thus been exempt from the germs, but no-one else would be. We are all only one step away from being eaten alive by bacteria, and other micro-pests. In the end, we all are, unless we are quickly incinerated.

Why would an all-powerful God have made a system that just looks exactly like a genetic experiment by an F-grade extra-terrestrial? You tell me. Biologically the world looks like a genetic mistake rather than an instance of divine creation. Intelligent design? Why, sure, but not necessarily by "God," probably from a science student from another universe, who failed, "Universe Creation 101."

Why bother with such questions? Because we are preparing for the collapse of Western civilization and a new reign of terror and chaos. We need to be right about our philosophies for the final battle. Otherwise, we just continue to embrace universalism, like a good host, and let ourselves get done over in another round, the final round.

---

[95]   https://www.amazon.com/Scars-Evolution-Elaine-Morgan/dp/019509431x.

[96]   https://en.wikipedia.org/wiki/Human_feces.

IV. CHRISTIANITY AND PEDOPHILIA

Here is a real problem for the Christians; kiddy fiddling. I will use an Australian example, since this is where this book is published, just for the Aussies so they do not feel neglected in the absurdist stakes, for after all, you gave us the "pearl."

The Australian Catholic Church has released "grim" data revealing 7 per cent of priests, working between 1950 and 2009, have been accused of child sex crimes.

The worst-offending institutions, by proportion of their religious staff, have been shown to be the orders of brothers, who often run schools and homes for the most vulnerable of children.

This is the most substantial dataset released to date about the extent of child sex abuse within the Australian Catholic Church, and was done with cooperation from them as part of the Royal Commission into Institutional Responses to Child Sexual Abuse.

The church surveyed 10 religious institutes and 75 church authorities to uncover the abuse data on priests, non-ordained brothers and sisters, and other church personnel who were employed between 1950 and 2009.

Counsel Gail Furness, SC, said 4,444 alleged child sex abuse incidents were recorded in the survey.

Ninety per cent of the victims were boys, with their average age at time of abuse being 11-and-a-half years old.

Girls were only 10-and-a-half years old on average when they were abused."[97]

Now even though my Christian critics would like to meet with me for a good old-fashioned US six shooter match at High Noon, can we not agree on this one, that it is horrific? I thought that this was just a problem localised to the Catholic Church because of celibacy, but there are claims from Protestant church leaders that the abuse issue is even worse in Protestant churches. How could that be?[98]

So, my dear enemies, how do you explain this utter corruption of church leadership going right to the very top?[99]

Sure, you have original sin and all that in your handy dandy tool kit, but really, since we are all immersed in it, original sin does not *specifically* explain the issue, does it? As well, the humanists complain about Christian prudery etc., vilifying Christian philosophers such as St Augustine, Tertullian etc., but there is no way that they would have engaged in the depravity of the moderns. They advocated exactly the opposite. Compared to the moderns, they really were saints, and deserved the title.

Nor am I saying that all Christians are pedophiles, for of course, almost all are not, and most highly value family and children, and are basically good people. I hate to even think this, but even my enemies here, at the end of the day, are probably good men too. But, your higher leaders seem to have joined the ranks of Satan, as you might put it in your paradigm. Why would God allow His Church to be destroyed? Free will? Hardly. And, what are you doing about this? Nothing.

---

[97] http://yournewswire.com/pope-francis-pedophile-priests-australia/; http://www.abc.net.au/news/2017-02-06/child-sex-abuse-royal-commission:-data-reveals-catholic-abuse/8243890.

[98] https://www.theguardian.com/commentisfree/2015/may/29/protestants-abuse-catholics-methodist-church; https://www.huffingtonpost.com/2013/10/01/protestant-sex-abuse-boz-tchividijian_n_4019347.html.

[99] https://nypost.com/2017/07/05/vatican-cops-bust-drug-fueled-gay-orgy-at-cardinals-apartment/.

I am at a loss to explain the preoccupation with kiddy fiddling by the Christian elites. I can see why the DC pizza eaters may play this power game, but why the Christians too? If it is just the expression of power, why doesn't God teach them a bitter lesson? Doesn't He answer prayers? What about the thousands of prayers by abused children, some traded in "priestly" sex gangs? It is an abomination. Any half-decent God should make their small dicks drop off.

On this issue alone, I am motivated to sweep the entire Christian trip into the dustbin of history, as most people are doing. No issue has turned people off Christianity like this one, so it is really a problem for you, at least sociologically. For me, it is more proof that your God simply does not exist.

PART II

BARBARIANISM

V. MANHOOD RISING

From Jack Donovan to James LaFond

Our post-industrial post-everything society is one where supreme value is consumption. It is one well suited for the success of women who are generally obedient workers and dutiful consumers. Thus, it is no surprise that *The Shriver Report* concludes that women are the new breadwinners, with women now comprising half of the US workforce.[100] It is a time of the twilight of manhood, the "decline of men" or even "the end of men" with feminists dancing on the supposedly grave of the "John Wayne man."[101] Indeed, according to this gleeful narrative, the "John Wayne man" has well and truly ridden off into the sunset, with boys now addicted to computer games and internet porn, incapable of fighting their way out of a wet paper bag.

Jack Donovan writing in his book *The Way of Men* stitches together the rise of women and global capitalist consumer civilization and the fall of manhood:

> Civilization comes at a cost of manliness. It comes at a cost of wildness, of risk of strife. It comes at a cost of strength, of courage, of mastery. It comes at a cost of honour. Increased civilization exacts a toll of virility, forcing manliness into further redoubts of vicariousness and abstraction. Civilization requires

---

[100] Maria Shriver, *The Shriver Report: A Woman's Nation Changes Everything*, The Centre for American Progress, October 16, 2009, at http://shriverreport.org/.

[101] Lionel Tiger, *The Decline of Males*, (Golden books, 1999); Michael Kimmel, *Guyland*, (HarperCollins, ebooks, 2008); Guy Garcia, *The Decline of Men*, (HarperCollins, ebooks, 2008); Hanna Rosin, *The End of Men: and The Rise of Women*, (Riverhead, 2012).

men to abandon their tribal gangs and submit to the will of one big institutionalised gang. Globalist civilization requires the abandonment of the gang narrative of us against them. It requires the abandonment of human scale integrity groups for "one world tribe." The same kind of man who once saw their own worth in the eyes of peers who they depended upon for survival will have to be satisfied with the "social security number" and the cheerfully manipulative assurances of their fellow drones. Feminist civilization requires the abandonment of patriarchy and brotherhood as men have known it since the beginning of time. The future being dreamed for us doesn't require the reimagining of masculinity, it ultimately demands the end of manhood and the soft embrace of personhood that has long been a feminist prescription for this ancient crisis of masculinity.

The emasculation of men[102] becomes complete with the metrosexual, an over-urbanised, over-consumerized shell of what was once a man, obsessing over body image, spending massive amounts of money on grooming and "looking hot."[103] The decline of manhood is tragically seen in the degrading of the value of physical strength, in part a product of the elimination by machines (to increase capitalist efficiency) of hard manual labor and physical lifting. Even those who visit gyms tend to cultivate looking good by high repetition workouts with light weights. The old school strongmen trained differently with low repetitions with heavy weights in compound exercises (e.g. squats, deadlifts, bench press, military press and Olympic lifts.) However, most men do neither manual work or gym work, working in an office at the computer. Combine that with little exercise and the modern high carb/high fat and salt junk diet and it is any wonder that men's guts have that pregnant look and their bodies are the consistency of jelly, jiggling as they thump away at computer keyboards. An epidemic of obesity, the product of a weak, anti-Darwinist, decadent civilization, is to be expected.

---

[102] Jack Donovan, "Everyone a Harlot," July 5, 2012 at https://www.boxingscene.com/forums/showthread.php?t=648050.

[103] Jack Donovan, Everyone a Harlot, above.

MEN'S RIGHT OR MEN'S WRONGS

What, if anything, is being done to oppose this genocide? Men's Rights activists, like feminists, want the dissolution of gender roles, along with economic, political and legal equality. They seek to abandon, with feminists, the age-old view of manhood associated with patriarchal societies of the past.[104] That, of course, is just the male version of feminism, and assumes that there is no intentional effort to destroy men, and that genuine men's rights will be granted by the globalist elites.

Pick Up Artists (PUAs), are primarily concerned with sexual conquests, of Second or Third World women. That approach of phallocentric obsession and narcissism, presents no solution to the problems that men face, only a temporary avoidance. Ignoring morality and honor, it assumes that one has the money to indulge in such a lifestyle (e.g. to travel and bang away) and that one has no other commitments, such as property, a business and a career, or for most of us, just a job. In any case all or most could not follow the "my brilliant cock" lifestyle, because the System still needs its male slaves and would move to quickly frustrate the attempts of those opting out. In general, too many of the Dissent Right, "manosphere" sites and books are locked into a pro-consumerist, materialist me-generation narcissism characterized by weakness and fear. No need to name anyone because these guys don't like criticism, and they are not our enemies, just young guys needing direction and manning-up. Fortunately, some look like they are already growing up and maturing, so hopefully in a few years, they will be doing useful work, rather than just looking at their dicks in the mirror of feminism. A few have already become strong political animals, fired up by the political chaos and have moved away from "how to get more snatch." I wish them well.

---

[104]   Roosh, "The Men's Rights Movement is Dead", at https://www.rooshv.com/the-mens-rights-movement-is-dead; Jack Donovan, "Long live the Manosphere", September 9, 2012, at https://web.archive.org/web/20130617174230/http://www.jack-donovan.com/axis/2012/09/long-live-the-manosphere; Paul Elam, "Adios, C-ya, Good-bye Man-o-Sphere," September 5, 2012 at https://avoiceformen.com/men/adios-man-o-sphere/.

THE WAY OF MEN

An alternative approach for men is represented by Jack Donovan's *The Way of Men* (2012).[105] A scholarly book dealing with the same problem of "What is a man", is Harvey C. Mansfield's *Manliness* published by Yale University Press.[106] Mansfield's book is for the intellectuals, those with an eye for literary detail. Donovan's book is addressed to ordinary men, and that is the source of its popularity.

Donovan begins with the basics. The way of men is the way of the gang or tribe. The core of masculinity is not seen as the hustler economics of capitalism, a recent historical abomination, but with a "small, embattled gang of men struggling to survive".[107] The gang or tribe stands to defend a territory, a behavior found in animals and for mammals, generally associated with males.[108] This territory defines the in-group and the out-group. As William Sumner said in *Folkways*: "Each group nourishes its own pride and vanity, boasts itself superior, exists in its own divinities, and looks with contempt on outsiders".[109] This is true today, except of course for Nordic (Northern Europeans), who have been most affected by the viral disease of globalism and cosmopolitanism.

The in-group is your tribe and the out-group other tribes which may or may not be a threat to your territory. Hobbes' war of all against all operates not at the individual level, but at the tribal level because humans have always lived in tribes, being social animals. Although Donovan views this biosocial fact positively, others who

---

[105]   Jack Donovan, *The Way of Men*, (Dissonant Hum, Milwaukee, Oregon), 2012.

[106]   Harvey C. Mansfield, *Manliness*, (Yale University Press, New Haven, 2006).

[107]   ibid p.3.

[108]   See K. Lorenz, *On Aggression*, Harcourt, Brace, New York, 1963, pp.34-39; E. O. Wilson, *Sociobiology*, Harvard University Press, Cambridge, 1975, pp.256-278. Mansfield in Manliness says: "lacking as women are, comparatively, in aggression and assertiveness, it is no surprise that men have ruled over all societies at almost all times." (p.64)

[109]   William Sumner, *Folkways: A Study of the Sociological Importance of Usages, Manners, Customs, Mores and Morals*, (Dover Publications, New York, 1959); Harold Robert Isaacs, *Idols of the Tribe: Group Identity and Political Change*, (Harvard University Press, Cambridge, 1975).

have written about territorial defense, view it negatively. Thus, Wrangham and Pederson in *Demonic Males* say: ""Demonic Males" gathering in small, self-perpetuating, self-aggrandising bands. They sight or invent an enemy "over there" – across the ridge, on the other side of a linguistic or social or political or ethnic or racial divide. The nature of the divide hardly seems to matter. What matters is the opportunity to engage in the vast and compelling drama of belonging to the gang, identifying the enemy, going on patrol, participating in the attack".[110] Mansfield also accepts that manliness "defines turf and fights for it, sometimes for no good reason, sometimes to defend precious rights."[111]

Men should be judged on possessing those qualities useful in guarding and defending territory. These qualities, Donovan argues, are tactical virtues. *Vir* is the Latin word for "man" and "virtue" comes from the Latin *virtus*, meaning manliness. For the ancient Romans, manliness meant martial valor.[112] As Roman civilization expanded, the concepts of virtues and manliness also expanded to include various civic and moral values, such as justice and honesty. However, these values vary more from culture-to-culture, and Donovan develops his account of manliness, the way of men, with reference to the fundamental fighting virtues, which are constants for men across history.

Jeff Costello[113] has argued that manhood/manliness may not be best viewed by looking at men in more positive and basic forms. Following Aristotle's argument about excellences, he believes that virtues should be looked at in the most developed form. Thus, Costello sees justice and honesty as manly virtues and he follows Schopenhauer and Weininger as seeing justice and honesty as virtues not strongly associated with women. Much depends upon

[110]   Richard Wrangham and Dale Peterson, *Demonic Males: Apes and The Origins of Human Violence*, (Horton Mifflin, New York 1996), p.248.

[111]   Harvey C. Mansfield, *Manliness*, p.20.

[112]   M. McDonnell, *Roman Manliness: Virtues and the Roman Republic*, (Cambridge University Press, Cambridge, 2006).

[113]   Jeff Costello, "Jack Donovan's *The Way of Men*," at https://www.counter-currents.com/2012/03/jack-donovans-the-way-of-men/.

definitions here, as women today certainly have contributed a force to be aware of with "social justice" and "feminist honesty."

However, it seems that once one moves outside of the warrior virtues as defining manhood, there are an explosion of other "intellectual" virtues. Plato saw four cardinal virtues – wisdom, justice, courage and moderation. Frank Miniter in *The Ultimate Man's Survival Guide*[114] recommends these virtues along with many others.[115] The seven virtues of Bushido are: courage, benevolence, etiquette, honesty, loyalty, rectitude and honor. But one can add to this list, of "higher" manly virtues. Some include:

1. independence - living by his own lights and following his own way, as he defines it;

2. authenticity - walking a man's talk, integrity (people know exactly where they stand);

3. freethinking and skepticism, a lack of mindless conformity;

4. valuing freedom and rebelling against tyranny;

5. pursuing the truth as it is perceived and

6. protecting kin and tribe and putting all of these values over money and material possessions.

Working along these lines one could develop a highly intellectualized account of manhood/manliness.

Nevertheless, I believe that Donovan is correct in focusing on the fighting virtues in developing this account of the way of men, qualities needed in survival situations. Such qualities include strength, courage, mastery and honor, territory protecting virtues. "Higher" virtues can only exist if these more fundamental, foundational virtues are present in sufficient numbers of men to protect the tribe. Further, these virtues are needed when human

---

[114] Frank Miniter, *The Ultimate Man's Survival Guide: Recovering the Lost Art of Manhood*, (Regnery Publishing, Washington DC, 2009), pp.175-179.

[115] ibid, p.116.

civilization collapses. As Tyler Durden says in the manhood movie *Fight Club* 1999:

> In the world I see – you're stalking elk through the damp canyon forests around the ruins of Rockefeller Centre. You will wear leather clothes that will last you the rest of your life. You will climb the wrist-thick kudzu vines that wrapped the Seas Tower. You will see these tiny figures pounding corn and lying strips of venison on the empty carpool line of the ruins of a superhighway.

THUMOS

Costello also refers to Plato's notion of *thumos*, which in the Republic is taken to be a core characteristic of the guardians of the Republic. Aristotle, as well, discusses *thumos* in the *Ethics* and other works.[116] There is considerable scholarly debate about the ancient Greek concept of *thumos*. Some see *thumos* as the spirited part of the soul, striving for honour and victory and recoiling against injury and injustice, an essential social phenomenon, while others see *thumos* as primarily concerned with self-preservation and survival.[117] Costello, however, sees *thumos* as the spiritual, transcendental part of us that strives to be more than merely human and stands behind the virtues. D. H. Lawrence, himself, unfortunately, not a good example of manhood, is quoted in this context as linking *thumos* and masculinity:

> It is the desire of the human male to build a world: not 'to build a world for you, dear,' but to build up out of his own self and his

---

[116]   W. F. R. Hardy, *Aristotle's Ethical Theory*, second edition, (Clarendon Press, Oxford, 1980).

[117]   S.M. Purviance, "Thumos and the Daring Soul: Craving Honour and Justice," *Journal of Ancient Philosophy*, volume 2, 2008, pp.1-16; H. A. Kraugerud, "'Essentially Social'? Discussion of the Spirited Part of the Soul in Plato," *European Journal of Philosophy*, vol.18, no.4, 2009, pp.481-494.

own belief and his own effort something wonderful. Not merely something useful. Something wonderful.

Nevertheless, there are alternative conceptions of *thumos* or spiritness and Donovan mentions one interpretation which sees *thumos* as a quality of aggression or drive, much like the "gameness" of the dog in dogfighting. Mixed martial arts fighter Sam Sheridan in *A Fighter's Heart*[118] refers to this "gameness" as "the eagerness to get into the fight, the berserker rage, and then the absolute commitment to the fight in the face of pain, of disfigurement, until death." Mansfield also characterizes *thumos* by comparison to the aggression of dogs: *thumos* is the "bristling snappiness of a dog ... As a dog defends its master, so the doggish part of the human soul defends the human ends higher than itself."[119]

Thumos in this sense should be added to Donovan's list of fighting virtues that define manliness. *Thumos* is more than mere courage, as it will be defined, and far more than aggression and assertiveness, which as Mansfield says, women are comparatively lacking in.[120] *Thumos* could be defined as the living flame behind the warrior spirit and a primal warrior value.[121] Richard Strozzi-Heckler says in his book *In Search of the Warrior Spirit* that the way of the warrior is not something to fearfully reject, but is rather something men should embrace:

> Part of being human is the longing, or perhaps even need, for the experiences of courage, selflessness, heroism, service and transcendence. Young men have traditionally been led to believe that war will provide the sole context for these experiences ...

---

[118] Sam Sheridan, *A Fighter's Heart: One Man's Journey Through the World of Fighting*, Grove Press, New York, 2007.

[119] Harvey C. Mansfield, *Manliness*, p.206.

[120] ibid, pp. 50-81 and p.64. The learned Professor says: "despite the laws, the customs, and the morals that we live under, it is still a considerable fact that almost any man can beat up almost any woman". (p.42). But, from the puny examples of computer game boys that we have seen, even this may no longer be true.

[121] John F. Gilbey, *The Way of a Warrior*, North Atlantic Books, Berkeley, 1982.

Instead of categorically disclaiming war as an evolutionary back fall into animal territoriality, or blaming industrialisation, technology, geographical boundaries, ideologies, or the modern state or burying our combative urges under exotic new age rituals, we urgently need to embrace our warrior impulse.[122]

MANLY VIRTUES

Strength, both physical and mental (determination) are key fighting virtues. Probably a better depiction is psycho-physical prowess. Men are on average physically stronger than women; the weak man is typically viewed as being less manly, although a woman is not considered less womanly if weak. Strength in men is the result of progressive resistance training, adequate nutrition and ample testosterone, the male sex hormone. Testosterone levels seems to be declining in men. Boosts can be made by medically prescribed injections administered by a doctor, or better yet naturally.[123] In a nutshell, men need to eat manly diets, of a rich variety of vegetables for vitamins, micronutrients and antioxidants, adequate levels of proteins and (good) fats and much, much less carbohydrates from sources such as bread and pasta. Fats are needed in the body for the mobilization of fat-soluble vitamins A, D, E and K, and also to trigger testosterone production. A mainly diet needs to be combined with, relative to one's level of fitness and age, relatively heavy weight training, and some form of fighting, with fists, legs, grappling, sticks and blades, just to keep the animal juices flowing.

The fighting virtue of courage is closely linked to warriorhood and manliness. As Carl von Clausewitz said in *On War*: "War is the province of danger, and therefore courage above all things is the first

---

[122] Richard Stozzi-Heckler, *In Search of the Warrior Spirit*, third edition, North Atlantic Books, Berkeley, 2003, pp.79-80.

[123] Two great books on naturally boosting testosterone levels are: Myatt Murphy, *Testosterone Transformation*, Rodale, New York, 2012 and Lou Schuler, *The Testosterone Advantage Plan*, Rodale, London, 2012.

quality of a warrior."[124] The Greek word *andreia*, meaning "courage", is derived from *andros* meaning "masculine." Aristotle, in Book III of the *Nicomachean Ethics*, gave an influential account of courage as seeing the courageous person as fearless in the face of death.[125] An alternative conception, as expressed by John Wayne is that manly men, even if afraid, "saddle up anyway" and go into battle regardless, facing pain and death for a higher good. Courage is linked to *thumos*, for it is the will to hang on, to keep struggling and fighting to the end – "true grit," to use another John Wayne reference. Such a man has gravitas or weight – he won't tolerate being pushed around and therefore deserves respect—the right not to be liked—but to be taken seriously. As John Wayne's character in his last movie *The Shootist* (1976) says:

> I won't be wronged, I won't be insulted, and I won't be laid a hand on. I don't do these things to other people, and require the same from them.

Honor is another fighting virtue. Honor is the value of a person as seen by society, the totality of the "traditions, stories and habits of thought of a particular society about the proper and improper uses of violence."[126] But, as Donovan notes, there is a much more basic type of honor, a right to respect, to be treated as having a certain value. Honor is a manly virtue; in societies where honor is important, the value is prescribed to men, and it requires an honor group to bestow honor. Donovan doesn't mention it, but loyalty is another manly virtue, closely linked to honor, but in some respects transcending it. Loyalty is the glue that holds tribes together, the commitment to defend the group come what may. Loyalty is a product of group identity and belonging – one is loyal because being so, defines who one is.

---

[124]   Carl von Clausewitz, *On War*, Routledge and Kegan Paul, London, 1949, p.47.

[125]   M. G. Brady, "The Fearlessness of Courage," *Southern Journal of Philosophy*, vol.43, 2005, pp.189-211.

[126]   James Bowman, *Honour: A History*, Encounter Books, New York, 2006, p.6.

Mastery ties together the manly virtues, being the proficiency and capability to use technologies and skills to control one's environment. Mastery and technics allow an increase group status for individuals having high skill levels and special skills. This virtue depends upon a sense of practical intelligence, not bullish learning, but a real-world ability in problem-solving and getting the job done.

BEYOND DONOVAN: LAFOND

The work of James LaFond goes beyond that of Donovan in the taboo-smashing stakes. His literary work resembles that of Louis-Ferdinand Céline (1894-1961), in its dark view of human nature and degeneracy. As he says in his philosophical masterpiece, *Taboo You: Way of the Terminal Man* (2016),

> While Jack wishes to save men from their yearning to be more like women, I say let that 70% die on the vine. Genghis Khan would have killed them all.(p. 130)

James is spot on in his natural selection culling thesis, as a garden which becomes too full of weeds, ultimately degenerates, degrading ecological capital. Too many weeds have artificially survived in the nanny welfare state, and the dark future ahead will sorely test their survival instincts.

The terminal man, unlike the tribal man of Donovan, is a loner, the man with no name in a world where the wild west is on every decaying street. Such a man, a neo-barbarian, would be the ultimate survivors at the end of battle for existence which we are now entering into.

CONCLUSION

In reviewing this material, it is clear that modern "man" is a long way from classical manhood. The feminists see this is showing that the "way of men" is over. I disagree; the age of the Y-chromosome is only just beginning. Our era of over-consumption, over-civilization and softness, is a singularity in history. It will not last. It is inevitable that the "way of men" and the taboo man will return. Those of us still capable of thinking and acting should adapt, or perish.

To conclude, we can do no better than to quote John Wayne, being in a John Wayne mood:

> I want to play a real man in all my films, and I define manhood simply: men should be tough, fair and courageous; never petty, never looking for a fight, but never backing down from one either.

And, "one day those doctrinaire liberals will wake up and find the pendulum has swung the other way." Yes, and they will be crushed by the earth-moving machinery of history.

VI. MIGHT IS RIGHT!

It's the Logic of Today

With the power elite transforming every square inch of the planet into a battleground, people of Northern European descent need their own warrior book. Why, everybody else seems to have theirs, so in the name of multiculturalism, let the inquiry begin.

Next to the LaFondian canon, may I recommend *Might is Right or Survival of the Fittest*, by Ragnar Redbeard. Just to warm up though, get some background music playing.

And, you could listen to this while driving your cement mixer, semi or Mad Max vehicle.

Here is Ragnar's signature poem, dripping testosterone:

Might was right when Caesar bled
Upon the stones of Rome,
Might was right when Genghis led
His hordes over Danube's foam,
And might was right when German troops
Poured down through Paris way,
It's the gospel of the ancient world
And the logic of today.

Behind all kings and presidents -
All government and law,
Are army-corps and cannoneers -

To hold the world in awe.
And sword-strong races own the earth,
And ride the conqueror's car -
And liberty has never been won
Except by deeds of war.

If Freddy Nietzsche had hair on his nut sack, this is exactly the sort of book he should have written, getting rid of the small amount of toxicity produced as an over-reaction to his critique of Christianity.[127] Nietzsche's madness from cancer, not from fucking a syphilitic whore? I am disappointed; I really liked the madness from syphilis explanation, instead of the McCancer one.

Naturally, Redbeard has had his critics. In the Loompanics version of the text S. E. Parker, gives an often-cited criticism in his introduction (pp. iv-vi). Redbeard, it is argued eschews moral codes, seeing them as, to use modern lingo, social constructions, what is "natural" is "right" and that is contradictory. Likewise, if all beings are "differentiated egos," then why should such beings accept the ideal of following natural laws? Parker, also mentions that nature is a "mental construct, not a fact," following the trendy sociology of today.

First, nature is not a mental construct, because the mind of human primates is just matter, as shown each day in slaughterhouses created by the same ideology which has given us social constructionism. If nature was a mental construct, then sacred social entities such as minorities and refugees are also mental constructs, and then could we not think them away?

The rejection of morality is the rejection of the traditional universalistic system of supposed justified propositions given to us by Christianity and liberalism, and is thus a form of moral nihilism,[128] that moral propositions are not justified, or true, as there are no moral facts. It is still possible to claim that living in accordance with

---

[127]   http://www.telegraph.co.uk/education/3313279/Madness-of-Nietzsche-was-cancer-not-syphilis.html.

[128]   https://en.wikipedia.org/wiki/Moral_nihilism.

natural principles, while not morally right, is still in one's biological interests, as it promotes survival value. Ragnar would not argue about whether or not survival was a "good" thing or not, because he rejected the ultra-rationalist masturbatory game of philosophy, requiring a reasoned justification for everything. In this respect, his position is more in tune with contemporary psychology, which has seen humans as limited in reasoning capacity, and more irrational than rational because of unending cognitive errors.[129]

So, Redbeard is not inconsistently saying that it is morally right to embrace the natural way, but instead is rejecting the conventional moral way of thinking, and only referring to utility as a guide for action.

Parker claims that Redbeard is inconsistent in rejecting moral codes, such as in Christianity and liberalism, but criticizes female promiscuity and sexual degeneracy "in a language redolent of the very Christian morality he so fiercely attacks elsewhere." (p. v) But, again, this is not a valid argument against him, because the mere shape of language does not determine its content and truth value. The actions of women are seen as having deleterious survival implications, which produce numerous harms, not moral harms, but physical ones. As such, the condemnation proceeds from there. Morality has nothing to do with it. Look, just think utilitarianism, but wipe away all of the moral talk. Replace it with "interests," and bingo, there you have it, consistency, and balls in one package. Thank you, expensive US university education that took forever to pay off.

Parker's master argument is that old Ragnar is a "racist," believing that his tribe of Anglo Saxons is superior, which would have been a common belief back in 1896, when the book was first published. Parker then, gloatingly proclaims that given the might is right philosophy, there can be no objection by Ragnar to non-Anglos grabbing power, which they have certainly done. That, however, is not an objection, since Ragnar would accept the implication. It does not follow that he would like it, or support it, and no doubt would urge his tribe to fight. But, if they do not, they will perish, as they

---

[129]   https://en.wikipedia.org/wiki/Cognitive_distortion.

now are. And, under present conditions, this existential threat exists totally regardless of all moral considerations.

Parker and others say that the fundamental contradiction in Redbeard's position is that there are no "rights" outside the "might" of the individual. That interpretation arises from taking passages of text out of context. Nothing prevents Redbeardians from embracing all the results of evolutionary biology, seeing human as tribal creatures, and putting a value on tribe survival, not a moral value, but a life value. It is hard for normies to operate in a scenario where their sacred universalistic values, and moral talk, are abandoned.

After writing that I came across this quote from N. Glazer and D. P. Moynihan, *Beyond the Melting Pot*, which may explain how people other than Nordics, accept particularism and reject universalism with no problems at all, with an attractive life as well:

> Edward C. Banfield has named the characteristic outlook of a small southern Italian village "amoral familism." According to this outlook, one owes nothing to anyone outside one's family, and effort should advance only the family ... the content of this moral code remains basically the same among Italian immigrants to America. One should not trust strangers, and may advance one's interest at the cost of strangers. Also, one does not interfere with strangers' business. One therefore tolerates the breaking of law by others (leaving aside the fact that it might be dangerous to do otherwise) ...

The basic idea of one tribe following might is right, and acting on it, while others surrounding them do not, and perish (or adopt might is right as well), is argued for by Andrew Schmookler, *The Parable of the Tribes* (1995).[130]

A more philosophical piece addressing Western civilization decline, along with the Faustian man theme is Duchesne's book,

---

[130] A. Schmookler, *The Parable of the Tribes*, 2nd edition, (State University of New York Press, Albany, 1995).

Faustian Man in a Multicultural Age,[131] but this book still has training wheels on it, believing that the system will survive. Still, a good read though.

---

[131] https://www.amazon.co.uk/Faustian-Man-Multicultural-Ricardo-Duchesne/dp/1910524840.

VII. WERE THE VIKINGS POLITICALLY CORRECT?

Professor Ricardo Duchesne, in one of his papers[132] notes that cultural Marxism has now deeply penetrated into the core of the somewhat harder sciences, having infected the flaccid sciences such as anthropology and sociology. The fields of population genetics, archaeology, paleogenetics, and evolutionary biology, are also following these soft "sciences." The aim of this penetration of intellectual corruption, whether the useful academic idiots are aware of it or not, is to aid programs such as the "Great Reset," and the "Great Replacement."

Professor Duchesne goes into greater detail than possible here, debunking the idea that the once hard sciences show that "there is no such thing as a European," because these people had a diverse origin over long time periods, from a mixture of multiple migrations. Of course, there is no such thing as an African or an Asian by the same argument. Likewise, the same vagueness argument can be applied to animal and plant species, as well, leading to a type of biological nihilism.

This deconstruction of European history, especially of the Nordic people, is seen clearly in present attempts to show that the Vikings were all very modern and politically correct, living in exactly the sort of contemporary society as today.

---

[132] "Deceptive Use of Scientific Data to Promote Ethnocide of Europeans," (October, 26, 2017), http://www.eurocanadian.ca/2017/deceptive-use-scientific-data-to-promote-ethnocide-europeans.html.

For example, Helen Buyniski, in her article "Revisionist Scholars Risk Reversing Decades of Women's Gains When They Declare an Unearthed Viking Woman Warrior is Transgender," August 21, 2020,[133] perhaps overstates her case, since most people have never heard of the discovery of a 10th century skeleton, a high-status Viking warrior, confirmed by DNA analysis to have had two X chromosomes.

Buyniski quoted Swedish archaeologist Professor Neil Price, who thought that this Viking, "may have been transgender … or non-binary, or gender fluid," backwardly projecting today's categories. Needless to say, there is no evidence at all for this.

Then there is Amy Jefford Franks who pushes the line that "Vikings were gay," and "there were queer Vikings," but admits that "it was not widespread." How does she know? She has a paper, "Odinn as a Queer Deity Mediating the Warrior Halls of Viking Age Scandinavia,"[134] which offers no real evidence for this claim.

Buyniski recognises that Viking society had rigid gender roles, so it is not unreasonable to suppose that the woman was not a warrior, but was just buried in those clothes, which certainly has occurred before in history, as women like to dress up, and someone may have dressed up the corpse. But, even if this was a female Viking Boudica type woman, it is possible that she was an exceptional person, which does not prove much, since statistical randomness in the universe will ensure that.

As pointed out by Robert Hampton, in his article, "Multicultural Vikings,"[135] the re-imagination of the past of Northern Europe, is solely performed to prevent the re-imagination of Northern Europe by white nationalists. As such, overtly political interpretations from both the Left and the Right are trashing actual historical facts.

Professor Dorothy Kim, in an op-ed article in *Time* magazine, says that the far Right is using Viking medievalism to create "narratives,"

---

[133]   https://www.rt.com/op-ed/498703-viking-warrior-woman-transgender-misogyny/.

[134]   https://www.academia.edu/40413584/valf%c7%aa%c3%90r_v%c7%aalur_and_valkyrjur_%c3%93%c3%90inn_as_a_queer_deity_mediating_the_warrior_halls_of_viking_age_scandinavia.

[135]   https://www.amren.com/commentary/2019/04/multicultural-vikings/.

VIKING AGE BARBARIAN

so the Left has to create "counter-narratives" as a show of force. And, Kim's counter-narrative is that the Vikings were multicultural. They were not homogeneous seafarers as is often imagined.

She sees the problems lying with the ideas expressed by the 19th century Romantic German nationalist Völkish movement, with the main sinners being the Brothers Grim, and Vilhelm Grønboch's *Vor Folke aet i Oldtiden* (*The Culture of the Teutons*. Some writers, such as Gustav Neckel, Bernhard Kummer, and Ottor Höfler did go down these Teutonic lines, but National Socialist ideology rejected "Odinism," in favour of Christianity, and many Odinists of the time were quickly marched off to concentration camps, since their anti-industrialism and anti-modernity was perceived as a threat. So, there is by no means some straight path from Teutonic Romanticism, to Nazi fanaticism. As noted by P. Jones and N. Pennick, *A History of Pagan Europe* (Routledge, London, 1995), "Hitler's rise to power came when the Catholic party supported the Nazis in the Reichstag in 1933, enabling Nazi seizure of power." (p. 218) Go blame it on the Christians!

But what really worries Kim is the notion of *Männebunde*, "all-male warrior associations in so-called primitive societies," and the "neo-pagan resurgence," in the forms of eco-fascism and Odinism. Eco-fascism does not have any link to Odinism, and is an extremist response to modern, largely Left-wing, environmentalism. Kim depicts Odinism as "masculinity based on the belief that the "barbaric" warriors of medieval Northern Europe functioned as a violent warrior *comitatus*." Well, if they were not violent warriors, then what was all the raiding and killing of Christians about, from Lindisfarne (793 AD), onwards? And contemporary Odinism has many forms, with female leaders, such as Freya Aswynn, which would have been easy enough to Google.

Another article along the same lines is by Clare Downham, Senior Lecture University of Liverpool.[136] Firstly, she says that the word "Viking" entered the English language in 1807 (old English had

---

[136] https://theconversation.com/vikings-were-never-the-pure-bred-master-race-white-supremacists-like-to-portray-84455.

85

wicing 300 years earlier), at the time of nationalism, and the Vikings were seen by the European colonists as "prototypes and ancestor figures." Downham does grant that the word "Viking" is "generally synonymous with Scandinavians from the ninth to the 11th centuries," and that they had a distinctive language, and culture. That should be more than enough to constitute an identity group, but apparently not for Northern Europeans. Thus, Downham says that Vikings were defined by an activity, "mobility," but much of the Scandinavian population did not "go-a-Viking." True, most were farmers, but in some seasons, some farmers did grab an axe and head off with the rest of the *Männebunde,* leaving wifey to hold the fort at home, sexist buggers that they were.

Downham says that the terms used to describe the Vikings, such as *"rus"* did not have an ethnic connotation. However, one of the many meanings of "rus" is red, which is a reference to the red hair of many Vikings.[137] However, other names were also used, such as *"pagani"* by Christians, and more generally, *"pirati,"* with an obvious meaning.

Downham claims that Viking mobility through use of the long boats, led to a "fusion of cultures with their ranks," as the Viking success arose from their ability "to embrace and adapt from a wide range of cultures whether that be the Christian-Irish in the west or the Muslims of the Assasid Caliphate in the east." Yes, the Vikings enslaved the Christian-Irish and sold them to the Muslims as slaves, especially women for the harems, with the fairest women being in hot demand as sex toys. Tens of thousands of Celtic and Anglo-Saxon men and women were sold to the Muslims by the Vikings.[138] The enslavement of European Christians is clear evidence that the Vikings of the time were never any sort of proto-white nationalist group, but were before conversion, an anti-Christian pagan people, who were tribal.

---

[137]   https://sonsofvikings.com/blogs/history/vikings-in-russia-the-rus-of-kiev-and-the-varangians.

[138]   Michael Wood, *In Search of the Dark Ages,* (Oxford, 1987).

It is true, as Downham notes, that Vikings were not just "one hitters," but that they stopped off at various places, sometimes to sit out strong stormy seasons at sea. And, it is also true that Vikings formed alliances with local peoples; why not, it made raiding easier? However, it is interesting to note this bit of "xenophobia": "Written accounts survive from Britain and Ireland condemning or seeking to prevent people from joining the Vikings." There is also the arguable position of F. J. Los, who saw the Viking raids on Christian parts of Europe as a revenge for the genocide committed by Christians against Germanic pagans.[139]

Then there is David Perry[140] who makes similar remarks to Downham. As Robert Hampton replies: "Perry at least admits the Vikings were predominantly Scandinavian and most of their cross-cultural contact was conquest and colonization. Yet he believes pillaging and enslavement were part of a rich cultural exchange."

It gets even more absurd. Funeral robes containing allegedly Arab inscriptions and Kufic script were discovered in 2017 in some Viking boat graves. This led some archaeologists to proclaim that Islam influenced Viking culture. The Vikings were pirates and stole everything that wasn't nailed down, and then burn that, stealing the nails. So why not suppose that some Viking simply stole the robes, by Occam's razor? In any case, Robert Hampton notes, some scholars hold that the robes contained no Arabic at all, and that Kufic script did not even exist in the Viking period. Other textile experts claim that the robes were simply not Islamic at all.

Juan Cole has said that genetic testing of some skeletons in Viking graves showed Iranian origins. So, "Iran may be the origin of a significant number of Vikings." No empirical evidence; mere speculation again. But now, Cole seems to be logically committed to accepting that there can be a genetic difference between Scandinavian Vikings and the alleged Iranian ones. And who, apart

---

[139] F. J. Los, *The Franks: A Critical Study in Christianisation and Imperialism*, (The Northern League, 1968).

[140] The Washington Post, https://www.washingtonpost.com/posteverything/wp/2017/05/31/white-supremacists-love-vikings-but-theyve-got-history-all-wrong/.

from Cole, says that the skeletons were Vikings anyway, given their "mobility" and passion for trade? The skeletons are more likely to have been slaves, and those of rank, dressed up in old abandoned rich guys' clothes, to confuse later generations archaeologists. It is just like if today people were buried with learned texts, like Aristotle, the texts being protected in the silent grave. Future archaeologists might come to think that these people were wise, when in fact, all they read was social media.

It is Cole's claims about Norse mythology which are the most absurd. Cole says that Norse mythology is a version of Indo-European mythology, so Thor, for example has equivalents found in Iran and India, which is true. But, the arrow of causation points the wrong way. In the pre-World War II period, it was widely held that this region was conquered by the Aryans, and surviving elements of their pagan warlord religion is found, in for example, the *Rig Veda*.[141]

Even given this though, common themes do not necessarily imply causation, since Norse mythology and Iranian and Indian myths could have produced this similarity by diffusion of ideas, or simply independently by common experience. A god battling a serpent is not a monopoly idea. In the Australian Aboriginal Dreamtime myths, snake battles as part of a creation myth occur as well, such as Lira and Kuniya. These are not derived from any other culture.

UNDERSTANDING NORDIC ORIGINS

I turn now to comments on a number of studies with relevant material on the question of Nordic/Northern European origins. Some of these studies have methodological limitations and flaws, but are still relevant to the present study. Unlike the material considered above, these studies are at least based within mainstream science, rather than postmodernist liberal arts faculties.

First, we consider, P. Rincon, "Stonehenge: DNA Reveals Origins of Builders," April 16, 2019.[142] This discusses, S. Brace (et al.), "Ancient

---

[141] https://www.civilseva.org/2018/02/advent-of-aryans-and-age-of-rig-veda.html.

[142] https://www.bbc.com/news/science-environment-47938188.

Genomes Indicate Population Replacement in Early Neolithic Britain," *Nature Ecology & Evolution*, vol. 3, 2019, pp. 765-771. Neolithic inhabitants who built Stonehenge, reached Britain about 4,000 BC. Their DNA resembled Neolithic peoples from Iberia, who originally migrated from the Mediterranean. They introduced building with megaliths. At the time, Britain was occupied by groups of "Western Hunter-Gatherers," but the groups did not interbreed much. These Western Hunter Gatherers were replaced by the Neolithic Farmers. As is now predictable, the academics claim that this genetic disappearance was not accountable by warfare, but due to "economic factors," such as "which lifestyles were best suited to exploit the landscape." That is incredible, since both groups had sufficient time to adapt to their environments, otherwise they would have been eliminated sooner. If the Western Hunter Gatherers did not know how to survive in their environment, they would have disappeared long before the stone draggers arrived on the scene. Thus, conflict is the more plausible explanation, but that is way too macho for today's political climate.

Burger (et al.), "Low Prevalence of Lactose Persistence in Bronze Age Europe Indicates Ongoing Strong Selection over the Last 3,000 Years," *Current Biology*, vol. 30, 2020, p. 1-9, discusses evidence of fourteen warriors from the Tollense Bronze Age battlefield in northern Germany, approximately 3,200 years BP. "Genetic data indicate that these individuals represent a single unstructured Central/Northern European population." (p. 1) Another sample was from 18 individuals from the Bronze Age site, Mokrin, in Serbia (~ 4,100 to ~ 3,700 BP, and 37 individuals from Eastern Europe and the Pontic-Caspian Steppe region (~ 5,980 to ~ 3,980 BP). All of three regions showed low lactose persistence, so that the "surge" in the allele causing lactose persistence among Eurasians, rs4988235-A, was not likely to be caused by expansion from the Steppes. As Brett Stevens notes, there was no "Eurasian invasion."[143]

---

[143] https://www.amerika.org/politics/periscope-september-26-2020.

Brian Sykes[144] reports that the drilling for DNA of a 12,000-year-old tooth found in the limestone caves of the Cheddar Gorge (not the famous Cheddar man, yet). His teams' DNA studies indicated that the ancestors of the native Europeans were hunter gatherers and not farmers from the Middle East. (p.10) The DNA of the Cheddar tooth was found to be virtually identical to modern Europeans, and this helped debunk the Middle East migration hypothesis, supporting a hunter gatherer ancestry for Europeans. Testing of the famous Cheddar Man DNA, found it to be "modern" too. (p. 11) One of the local history teachers, was a close DNA match. (p.11)[145]

Now back to the Vikings. An article by A. Margaryan (et al.)[146] allegedly "rewrote the history books", yet again, this is something all scientific papers in this field somehow are said to do.[147] One even claimed there were significant gene flows from Southern Europe and Asia.[148] The research involved the sequencing of the genomes of 442 humans from archaeological sites across Europe and Greenland, dating from 2,400 BC to 1600 AD, the later date being long after the Viking Age, which arguably ended in 1066 AD. But, early in the paper we get the motivation: "The Viking diaspora was characterized by substantial transregional engagement: distinct populations influenced the genomic make ups of different regions of Europe, and Scandinavia experienced increased contact with the rest of the continent." (p. 390)

They found that some of the skeletons were not of Scandinavian genetic ancestry, which commits them to the existence of a Scandinavian ancestry to begin with. (p. 392) Two individuals buried in Scandinavia were genetically similar to the Irish/Scots of today. (p.

---

[144] *Saxons, Vikings and Celts: The Genetic Roots of Britain and Ireland*, (W. W. Norton, New York, 2006).

[145] https://www.dailymail.co.uk/news/article-5364983/Retired-history-teacher-believes-looks-like-Cheddar-Man.html.

[146] "Population Genomics of the Viking World," *Nature*, vol. 585, 2020, pp. 390-396.

[147] https://www.invers.com/science/viking-dna-study.

[148] https://www.sciencedaily.com/releases/2020/09/200916113544.htm.

392) And, two others from Orkney, were 50 percent Scandinavian. (p. 393). We are not told about the Chinese Viking hybrids, sadly! And with regards to the Greenland Viking settlements, "sexual interaction between the Greenland Norse populations and these other groups was absent, or occurred on a very small scale." (p. 393)

Let us turn now to a methodological critique of the study. Scientific studies are only as good as their data, and in a study like this one, the sample needs to be representative to establish valid conclusions about the population. Thus, the study's data is 442 human skeletons from various sites across Europe and Greenland. That is a vast territory to begin with. But, what makes things even worse for the statistical validity of the study, is that the time period considered is from the Bronze Age (2400 BC) to the Early modern period (1600 AD), with the actual Viking Age being only a small part of that period. And, the period is 4,000 years! So we are to take 442 individuals over 4,000 years, some of them, if not most of them, not appearing in the Viking Age, and make genetic conclusions about the Viking Age ... go figure! Surely over that long time period there would be at least 442 visitors, slaves or travelers, who one way or another ended up buried in the graves that were found?

VIII. REVENGE!

There is nothing better, sweeter, in this world than getting revenge upon one's enemies, destroying them, utterly, today within the bounds of the law; after the collapse, by any means necessary. In a real world, not this mass produced, consumer-masturbatory one, a warrior should be free to live as Genghis Khan proposed in a disputed aphorism:

> The real greatest pleasure of men is to repress rebels and defeat enemies, to exterminate them and grab everything they have; to see their married women crying, to ride on their steeds with smooth backs, to treat their beautiful queens and concubines as pyjamas and pillows, to stare and kiss their rose-colored faces and to suck their sweet lips of nipple-colored.

Here is Arnold Schwarzenegger, saying the same, as Conan the Barbarian.[149]

Warlord: "What is best in life?"

Conan: "To crush your enemies, to see them driven before you, and to hear the lamentations of their women."

Then we have in another corner, the philosophers. I did not find much about revenge, even though restitution is an important part of the ethics of punishment. However, I found an interesting piece: G. Bar-Elli and D. Held (The Hebrew University), "Can Revenge be Just or Otherwise Justified?" *Theoria*, April 1986.[150]

---

[149] https://www.youtube.com/watch?v=Oo9buo9Mtos.

[150] https://doi.org/10.1111/j.2567.1986-1755.tb00100.x.

The paper is a technical piece in analytic philosophy, which means that it is impossible for our kind to try to read it in its entirety without falling into a restless slumber, punctured by nightmares, but I broke the spell by going backwards, and read the conclusion first, while chanting the name "Kant, Kant, Kant," as I did, really quick. Try it. Why work hard when one can cheat?

> We are led to the conclusion that the paradox of revenge lies not just in its incompatibility with our idea of justice but more deeply in the inner contradiction that it displays on the emotional level of human psychology. In other words, even if justice were to be restructured so as ideally to deal with all aspects of restoration and redress, it still could not formulate rules and principles for the satisfaction of that insatiable human thirst for revenge. The desire to take revenge may be justified by the incapacity of the legal system of justice to fully restore the previous situation; but we cannot appeal to justice for help; only for condonation. Revenge can never be part of the system of justice; nor can it be justified as 'just'. This does not mean, however that revenge cannot be morally justified. It may be regarded as morally deserved by the special kind of personal relationship in the particular situation. But as long as we are living in this world, a world in which the conventional system of social justice governs behavior, there will always be an unresolved tension between the practice of revenge and the demands of justice. The moral legitimacy of revenge can never be admitted by social justice. But since it has only limited purposes and scope, it is not surprising that justice is not always compatible with moral justifiability.

Translated: a social theory of justice will not capture cases where people fall through the cracks of its structure and do not obtain justice, but where there is a moral legitimacy for revenge. Even mundane mortals must experience this when viewing revenge flicks like *Death Wish* (2018), where Dr Paul Kersey turns into a vigilante because the criminal justice system is too overloaded to deliver justice in the case of the murder of his wife, and attempted murder of his daughter. There is a scene from another great revenge flick, *Instant Death* (2017) with Big Lou, someone I have always liked more than Arnie, whose screen daughter, after brutal dual-fucking, gets one of the worst beatings dished out short of death, with eyes stabbed. This low budget indie movie is worth looking at one rainy Saturday night, while you caress your AR 15 and its family of puppies, tipping down a beer and chewing beef jerky. So what if the movie is full of flaws; what isn't?

Anyway, in short, it has been argued that modern legal system arose to deal with the social problem of blood feuds and the quest for revenge. This seems to me to be yet another flaw of civilization, attempting to chain a legitimate human emotion that should never be imprisoned, for the sake of economics and fucking trade.

We therefore need this: Fuck Civilization, Hard/Revenge: Nothing is Sweeter, with capitals added for emphasis, put on T-shirts.

IX. CONFRONTING COLLAPSE

The Philosophy of Survivalism 101

Not all men seek rest and peace; some are born with the spirit of the storm in their blood, restless harbingers of violence and bloodshed, knowing no other path.

- Robert E. Howard[151]

INTRODUCTION

In this part I offer what I take to be the only viable response to the converging and compounding catastrophes discussed above. There are, of course, many lengthy responses that could be given, but here I will keep to basics and essentials, reducing the discourse to one discussion rather than a spiralling book. The following material is intended as only an introduction and not the last word on a vast topic, perhaps the most important topic remaining for humanity, or at least, for those remnants wanting to survive: survivalism.

The need to embrace survivalism, and hard survivalism at that, with a focus upon weaponry, will be mandatory for those who strive to survive the collapse and engage in the struggle for life in post-apocalyptic times. A "mere" economic collapse is likely to kill 25-50 million Americans in the first 90 days, from starvation, disease, and murder, and what has been presented here

---

[151]   Robert E. Howard, "A Witch Shall be Born," *Weird Tales* (1934).

is a wide-ranging cascading collapse, of even greater severity.[152] Of particular concern for those who choose to stay in cities will be the outbreak of disease when the shit not only hits the fan, but floods into the water supply, so that there will be an instant return with a vengeance of old favorite diseases such as typhoid, cholera, dysentery, cryptosporidiosis, schistosomiasis, polio, hepatitis A … from lack of sanitation.[153] Higher temperatures in some regions, if combined with wet conditions, may result in a ready over-supply of mosquitoes, carrying tradition diseases such as malaria and yellow fever, but now also western equine encephalitis and eastern equine encephalitis, zika, chikungunya, and dengue.[154] In the sort of SHTF scenario envisaged here, cities will become fetid morgues.

Apart from the disease aspect, social breakdown will ring in a hyper-violent world as people revert back to the Hobbesian state of nature due to the shortage of resources.[155] Much of the prepper and survivalist literature is not geared towards such possible near-term extinction threats, often addressing economic collapse situations, rather than a systematic undermining of the biological and ecological life support systems of the planet. It may well be that there is no ultimate solution beyond escape from the planet itself, as the late Professor Stephen Hawking proposed,[156] but that is not an option for us ordinary people, even if it was technologically viable.

Thus, we are left with the survivalist/prepper material on hand to face what is coming. In this part I will focus primarily upon the individual physical transformation needed to have a fighting chance.

---

[152]   J. D. Heyes, "Economic Collapse in the USA to Cause 25-50 Million Deaths in the First 90 Days … From Starvation, Rioting, Murder and More," June 10, 2016, at http://collapsenews/2016-06-10-economic-collapse-in-the-usa-to-cause-25-50-million-dead-.

[153]   http://www.collapse.news/2018-11-28-when-shtf-america-will-see-a-resurgence-of-disease.html.

[154]   http://www.survivaldan101.com/shtf-diseases-long-eradicated-developed-world-will-rear-ugly-heads-many-unprepared-will-die/.

[155]   https://www.naturalnews.com/2019-01-02-preview-of-the-chaos-to-come-in-2019-and-2020-if-you-dare-to-glimpse-reality.html.

[156]   https://www.businessinsider.com.au/stephen-hawking-humans-leave-earth-or-be-annihilated-2018-10?r=US&IR=T.

I am not discounting community/tribal survival strategies, but these have been discussed by others, and there is not much I can add.[157]

To get into the spirit of things, here is an inspired article for newbies to read: Daisy Luther, "Here's How You'll Die When SHTF (and How to Prevent Your Untimely Demise)."[158]

Most prepper/survivalist literature which accepts some type of catastrophic collapse, recommends a retreat away from the cities to either small country towns, or to homesteads, where families can create what Piero San Giorgio calls in his comprehensive overview book, a "Sustainable Autonomous Base,"[159] which is something of a generalization of leading US survivalist James Wesley Rawles' notion of the "American Redoubt."[160] Rawles and many other, including Boston T. Party,[161] have advocated for Americans who want to survive, moving to the mountain states of Idaho, Montana, Wyoming, East Oregon and Eastern Washington, for both agricultural sustainability and defense. In general, this movement essentially builds on both the 1960s "back-to-the-land" and nuclear retreatism movements. The core idea is for "rootedness, autonomy, permanence," based upon water, food, hygiene and health, energy, knowledge, defense and the social bond/community. San Giorgio's book is an outstanding introduction to people new to all of this, which I take to be who I am addressing now, since hardened survivalists would not be reading this sort of book, but prepping anyway. So, there is your first book to acquire, and the Kindle version is good and will save valuable dollars, although I hope you do support this good guy because few books in this genre offer so much. I will suggest a number of books needed as reference texts to keep for the dark times to come, but

---

[157]   https://www.youtube.com/watch?v=kVWx_Dqod_Q;https://www.youtube.com/watch?v=MoGAUeMz6fM.

[158]   https://www.theorganicprepper.com/how-to-die-when-the-shtf/.

[159]   P. San Giorgio, *Survive the Economic Collapse: A Practical Guide*, (Radix/Washington Summit Publishers, Whitefish, 2013).

[160]   https://www.conservapedia.com/American_Redoubt; https://survivalblog.com/redoubt/.

[161]   Boston T. Party, *Boston's Gun Bible*, (Javelin Press, 2002).

many should be read now and digested, and the vast majority can be got on the interlibrary loans systems of your countries, if like me, you are cash-challenged. In fact, this service can get overseas books, so there is no limit to knowledge acquisition by the sincere.

At the present time it is well worthwhile spending all free time educating and training for what is to come, rather than wasting time on pointless Hollywood mind-numbing "entertainment." One subject matter to begin acquiring books and internet literature on is that of homesteading, "country wisdom" and food self-sufficiency. *Mother Earth News*[162] and *Earth Garden*[163] give good introductions to most topics. One can then use the internet to search out further information, such as how to raise, say, meat-line chickens and protect them from predators etc. For convenience here are some books to consider looking at to get your feet moving, and I put these as in-text references for convenience: Simon Dawson, *The Self-Sufficiency Bible*, (Watkins Publishing, London, 2013); Editors of Storey Books, *Country Wisdom and Know-How: Everything You Need to Know to Live Off the Land*, (Black Dog and Leventhal Publishers, New York, 2004); Carla Emery, *The Encyclopedia of Country Living*, (Sasquatch Books, Seattle, 2012); Abigail R. Gehring (ed.), *Back to Basics: A Complete Guide to Traditional Skills*, 4th edition, (Skyhorse Publishing, New York, 2014); Abigail R. Gehring (ed.), *The Homesteading Handbook*, (Skyhorse Publishing, New York, 2011), and no such list would be satisfactory without John Seymour, *The New Complete Book of Self-Sufficiency*, (Dorling Kindersley, London, 2009), with its magnificent artwork, and foreword by *Small is Beautiful* author, E. F. Schumacher. These will serve as a good foundation for beginning survivalists to conduct their own education while the internet is up and running.

The vast majority of people are not going to have a retreat or a Sustainable Autonomous Base; indeed, people will be struggling to pay off the homes that they have, or trapped in the rent noose. That does not mean that when the time comes it will not be possible to "bug out" to some remote locality, which you simply "settle." I am

---

[162]   www.motherearthnews.com.

[163]   www.earthgarden.com.au.

not implying taking someone's property, which is an invitation for being shot, but to find suitable sustainable land, with water, where you, your family, and maybe a tribal network of families could settle. People will die of fright and commit suicide when the system collapses, so there will be some space created for those who can hang on. It is not contemplated that millions of people will even attempt to retreat, since we are supposing that the vast majority of people are going to die and rot, largely *in situ*. It is hardly unreasonable to suppose that some diehards could hold out, and start anew. After all, there are cases where lone humans have been raised by wolves, as in the case of Marcos Rodríguez Pantoja, who lived alone in the Sierra Morena mountain range in southern Spain with wolves, after he was abandoned at age 7 in 1953.[164] Further, there could well be better opportunities outside your present country, if it is a shithole: Fernando Aguirre, *Bugging Out and Relocating: What to Do When Staying is Not an Option*, (2014), is a good book on this topic, clear and no bullshit.

People lacking a Sustainable Autonomous Base, will need to bug out and find one, hopefully some locality that has been found prior to collapse. Survival chances are therefore lower than if one was living in one's rural retreat, but that is just as it is going to be for most people. Hence, the issue of bugging out, Get(ting) Out of Dodge (GOOD), is going to be of vital importance for most people. That means that would-be survivors are going to need to also beef up on issues associated with bugging out. In general, the idea is to get the hell out before the shit hits the fan, and thus, hopefully some large vehicle will be available, a van or truck, perhaps with a trailer. If one has this, then one can work out what needs to be taken, in your modern version of the pioneer's covered wagon ("prairie schooner"); if they did it, so can you, and more easily given the vast quantity of consumer goods available. To organize your stuff, it would be good to look at John Wesley, Rawles, *How to Survive the End of the World as We Know It*, (Penguin, New York, 2009), which introduces the meta-organizational principle of having a list of lists. Basically, the

---

[164] https://www.theguardian.com/news/2018/aug/28/how-to-be-human-the-man-who-was-raised-by-wolves.

idea is to put into what space you have, apart from spare fuel (who knows what could happen to availability), (1) weapons (firearms, ammunition, melee weapons); (2) clothing to last the years which is adequate for the weather conditions one will face, or may have to face if fleeing Mutant Zombie Bikers (e.g. both cold weather and hot weather clothes), and a supply of strong, suitable footwear to last the years; (3) water to last the trip, or at least until some can be found and water purification systems (even if just by filtering and boiling, and hence a large pot, and numerous fire starting methods from waterproof matches to flint and carbon steel strikers/magnesium fire starters, tinder (cotton balls soaked in petroleum jelly), so fire will be able to be made in the worst conditions); (4) a stockpile of long lasting food, as much as can reasonably be taken, consistent with having other essential items; (5), some sort of sleeping system, including sleeping mats, and a synthetic sleeping bag, adequate for the coldest climate you are likely to face; (6) a large supply of non-hybrid seeds of nutritious vegetables that will readily grow in the area that you are going to; (7) various hand tools for the garden, and for building a shelter, such as hand saws, that can be resharpened, at least one axe, hammers, nails, files (to sharpen blades and cutting tools) machetes and as much hardware items such as duct tape that can be reasonably packed; (8) medical and first aid supplies, especially antiseptics (Betadine, Mupirocin, band aids, bandages dressings etc., dental care products, painkillers etc.); (9) a sewing kit for clothes repair; (10) kitchen utensils, knives, forks, spoons, bowls, cups, plates, that are not breakable, cooking pot and fry pan, just to name some basic items; (11) odds and ends such as paracord, and any items that can be fitted in the remaining space that you now need to get by (non-electrical, perhaps grooming products, comb, hair cutting scissors etc.) fishing gear (assorted hooks, swivels, shot, fishing line), snares, snare wire and trapping gear, space blankets, tarps, raingear, hats, sunglasses, extra-large garbage bags, a tent, and so on, until the available space is gone.

Essentially the car evacuation scenario will encompass the standard, on-your-back bug out bag, spread over the space of the vehicle, and includes the bigger items that one could not practically be taken on one's back, or in a small cart, such as the garden tools. The articles "75 Essential Items for Bug Out Bag Packing List"[165] and "100 Items to Disappear First After the SHTF,"[166] are very helpful. They should be printed off as guides. To take things even further, consult Mark Puhaly and Joel Stevens, *Everyday Survival Kits: Exactly What You Need for Constant Preparedness*, (Living Ready Books, Iola, 2014), which has an ultimate bug out bag (pp. 149-162), that could be useful for TEOTWAWKI, and John McCann, *Build the Perfect Survival Kit*, (Krause Publications, Iola, 2005), is excellent.

The situation of bugging out on foot is one of the worst-case scenarios, but it could happen, even if one leaves by vehicle. Roads could be blocked, or one may need to abandon the vehicle if attacked, taking emergency gear, prepared earlier. This scenario reduces to the backpacker's trip from hell, and it would be well worthwhile to gain backpacking knowledge before TSHTF. The unquestionably best book on this is Chris Townsend, *The Backpacker's Handbook*, 4th edition, (McGraw Hill, New York, 2012). This is still the bible for backpackers, and has information to prepare one for foot trips lasting months. If you cannot find a new home by then, then you will probably die anyway. The book covers in detail footwear, the best types of backpacks, clothing, shelter food and water, and safety issues (e.g. wild animals).

Some notes made from Townsend's book: don't go cheap on footwear for walking, you need boots that are designed for treks, with protection of your feet and long-term comfort. Work boots, and runners will not deliver over long trips. This is like tyres for your car. Socks are important as well, and need to wick moisture, to prevent tinea and other foot problems; merino wool socks are good. Backpacks, if poor quality, will over a long journey of weeks, kill your back, so it is worth spending some hard-earnt dollars to get a large heavy-duty pack, with a good fit, a frame that puts the weight on the hips, and has an overall good suspension system. I am no expert on brand choice, so consult Townsend.

---

[165] https://bugoutbagacademy.com/free-bug-out-bag-list/.

[166] https://urbansurvivalsite.com/100-items-to-disappear-first-in-a-panic/.

Clothing; the layer system for cold weather has now become standard, rather than just one big, thick coat. In a nutshell, there is an inner layer of underwear to wick moisture (but not cotton, unless in the heat, as it will become soaked, fine merino wool is good): a thicker mid-layer for insulation, and an outer shell which is breathable, waterproof and windproof. In colder weather, probably a suitable parker is need, along with waterproof, windproof pants. For cold weather use: wool or fur hats, with a leather or synthetic covering to deal with rain. Shelter: a lightweight tent with fabric which is both waterproof and breathable (Gore-Tex, Cuben Fiber, eVent), otherwise you will wake up at 3 a.m. with water dripping down your nose from the roof of the tent. There are other options such as the Australian bush swag, which is a waterproof canvas sleeping "bed roll," usually insect-proof, and with a built-in foam mattress. The swag is lightweight, but still a bit heavy for most people to take on foot. As well, one is gift-wrapped in a confined space, which may not be the best thing to do in the fields of North America, compared to the relative safety of Australia. Those going ultra-light may opt for a Bivouac bag, also waterproof and breathable. This would require a sleeping pad or air mattress or self-inflating pad. Sleeping bags should be good quality synthetics, which usually can still keep one warm when wet (up to a point), for down bags will certainly lose their insulating qualities when wet.

It would be a good idea to take up backpacking now, before the collapse, to get a taste of what life in the great outdoors is like, to experience its pitfalls. That said though, it is one of the good things in life that should be enjoyed now before our descent into hell. Hence, one can have fun, while preparing for the collapse, and Getting Out of Dodge.

Closely associated with backpacking is wild country survival, the sort of thing which Bear Grylls and Les Stroud have made damn entertaining TV shows from. There is a serious literature on these bushcraft survival skills, such as making fire, purifying water, trapping, fishing, and building shelters. This material is covered in a number of superb books, the granddaddy of them all being by former British SAS soldier, John Wiseman, *SAS Survival Handbook:*

The Ultimate Guide to Surviving Anywhere, (Collins, London, 2009). This book will bring you up to speed on bushcraft survival skills, such as fire making, and purification of water, two absolutely essential skills needed for bugging out survival. My only point of disagreement is the claim on page 49, that a normal diet includes 10 g daily of salt, if I read this correctly; way too much, with the American Heart Association recommending 2.3 g a day as the limit, and 1.5 g being better.[167] Otherwise an excellent book. Along the same lines is Chris McNab, *SAS and Elite Forces Guide: Wilderness Survival*, (Amber Books, London, 2011).

Very detailed wilderness survival essays are compiled in A. Rost, *Survival Wisdom and Know-How: Everything You Need to Know to Subsist in the Wilderness*, (Black Dog and Leventhal Publishers, New York, 2007). With a North American focus, topics covered are food, hunting and fishing, drinking water, fire, shelter, travel on land, travel on water, navigation, dealing with weather and climate, first aid and much more. There is even an informative discussion of black and brown bears. Overall, 480 pages in a large format, double-column book, complete with a very good index.

There are two books taking a "minimalist" approach to survival, doing it with just the clothes on one's back and knife on one's belt. Of course, if there was a choice, no sane person would go this way, but gear can be lost or stolen, so you could wind up in this situation and thus should be ready. See: Bob Holtzman, *Adventure Survival Handbook: How to Stay Alive in the Wild with Just a Blade and Your Wits*, (New Burlington Books, London, 2012) and M. Elbroch and M. Pewtherer, *Wilderness Survival: Living Off the Land with the Clothes on Your Back and Knife in Your Belt*, (Ragged Mountain/McGraw-Hill, Camden, 2006).

Obtaining safe water while travelling could be the most challenging problem, next to self-protection. Typical water sources may be highly contaminated, perhaps with floating, badly decomposed corpses. Some very good articles on water treatment

---

[167]   https://www.heart.org/en/healthy-living/healthy-eating/eat-smart/sodium/how-much-sodium-should-i-eat-per-day.

are: AJS, "Water Treatment Options: How to Avoid Poisoning from Toxins – Part I"[168] and Part II.[169] These articles, written by a PhD scientist, are the most insightful that I have seen about the difficulty of obtaining uncontaminated water in a grid down, bug out situation. It offers a strong case for bugging out more quickly, but as I have recognized, most people will not be able to do this. Hence, some serious research time needs to be devoted to the water purification issue.

One of the key areas that would-be survivors should start educating themselves about immediately, is survival and collapse medicine. I am well aware of the standard prepper wisdom of having a doctor, dentist, or at worse, a nurse, in one's survival group, but seriously, how many people know medical professions well enough to invite them into the group, if they are interested at all? Having worked "around" the medical field, in security, I can say that most doctors, let alone surgeons, care only for money, and at a personal level, would probably die off quick in a grid down situation anyway. Obviously, if you have a good doctor, well, good luck to you. But most people are going to be on their own regarding health and medicine.

Thus, individuals need to take responsibility for their health immediately, not merely attempting to achieve excellent levels of fitness, with no smoking, drugs or alcohol at all, and with the diet described in this book, but also by educating themselves about medical matters. One can start doing this by researching a particular medical problem that comes along. Practice diagnosing the problem, read up on it from reliable medical sources on the internet and medical textbooks from the library, and have detailed knowledge before going to a GP. After a while you will begin to get a feel for diagnosing your medical problems, and not get taken for a sucker.

This knowledge should be enhanced by detailed reading. The iconic beginner's texts are David Werner, *Where There is No Doctor*,

---

[168] https://survivalblog.com/water-treatment-options-avoid-poisoning-toxins-part-1-ajs/.

[169] https://survivalblog.com/water-treatment-options-avoid-poisoning-toxins-part-2-ajs/.

(2017).[170] There are probably still old PDFs lurking on the net, but the 2017 edition is substantially updated. Along with this, dental problems are very important as well. The introductory text is Murray Dickson's, *Where There is No Dentist*, (2018).[171] Again, there are older versions available as PDFs, but you need the new one, and you need a hardcopy of this. In general, even if no other books recommended here are purchased (and this reading can be done using local library interlibrary loans), having hard copies of medical books could be the difference between life and death for you and your family. Speaking of families, child birth is usually neglected by preppers, so if you plan on dipping your wick, have on hand, Susan Klein (et al.), *A Book for Midwives* (2013),[172] as part of your library: These books were all written for people in the Third World, but a collapse situation will capture those levels of poverty and degradation, and beyond, so the books will be relevant.

Regarding collapse medicine, there are numerous prepper survival medicine books on Kindle, but most of these, while having some merit, are too superficial to put much faith in. There is a free down loadable 614 page book worth printing off: *Survival and Austere Medicine: An Introduction*, 3rd edition, (2017), by the Remote, Austere, Wilderness and Third World Medicine Discussion Board Moderators, dealing with grid down medicine.[173] The book discusses topics not dealt with in depth in other survival medicine books, including nursing care in a grid down situation, and dealing with the health threat of nuclear, biological and chemical warfare.

On collapse medicine, Joseph Alton and Amy Alton, *The Survival Medicine Handbook: A Guide for When Help is Not on the Way*, (Doom and Bloom, 2013), is excellent, and a paper version needs to be purchased for the medical reference library. The approach is primarily orthodox medicine, but more traditional herbal and

---

[170]   https://store.hesperian.org/prod/Where_There_Is_No_Doctor.html.

[171]   https://store.hesperian.org/prod/Where_There_Is_No_Dentist.html.

[172]   https://www.amazon.com.au/Book-Midwives-Pregnancy-Womens-Health/dp/0942364236.

[173]   https://survivalforum.survivalmagazine.org/forum/wilderness-survival-camping/wilderness-medicine/212213-new-edition-survival-austere-medicine-3rd-edition.

complementary treatments are mentioned where relevant. The book has a good index, and a survivalist would do well to work through the book, using it as a basis for one's own research. This can be supplemented with information at the Doom and Bloom website.[174]

Ralph La Guardia, *The Doomsday Book of Medicine: What Will You Do When There are No Doctors or Medicine?* (Mindstir Media, New Hampshire, 2015), is a similar book to *The Survival Medicine Handbook*, but with a more complimentary medical orientation, since La Guardia sees a grid down situation as leading to no doctors, and no conventional medicine, so that for example, honey might be used for wound healing. Conventional antibiotics are going to run out no matter how much is stored, because they have a shelf life. Thus, research renewable medical resources, such as herbal antibiotics: see Stephen Harrod Buhner, *Herbal Antibiotics: Natural Alternatives for Treating Drug-Resistant Bacteria*, (Storey Publishing, North Adams, 2012), which is extensively referenced. As well, get some good texts on herbal medicines: Alfred Vogel, *The Nature Doctor: A Manual of Traditional and Complimentary Medicine*, (Keats Publishing, 1991), David Hoffman, *Medical Herbalism*, (Healing Arts Press, 2003), and Rosalee De La Foret, *Alchemy of Herbs*, (Hay House, 2017), are good starting points. Natural news.com, has numerous articles on this topic, so search it using their search index.

I like Gerard S. Doyle, *When There is No Doctor: Preventive and Emergency Healthcare in Challenging Times*, (Process Media, Port Townsend, 2010), and James Hubbard, *The Survival Doctor's Complete Handbook: What to Do When Help is Not on the Way*, (Reader's Digest, New York, 2016), which, while not being as comprehensive as the other books listed above, do try to get you to think in a problem-solving medical diagnostic way, and getting a taste of medical methodology and thinking in a problem-solving fashion, is just as important as having a mass of medical facts at hand. Facts need to be ordered by theories to be of much use. While not collapse medicine as such, these two books are also useful: William Forgey, *Wilderness Medicine: Beyond First Aid*, 6th edition,

(Falcon Guides, Guilford, 2012), and Hugh Coffee, *Ditch Medicine: Advanced Field Procedures for Emergencies*, (Paladin Press, Boulder, Colorado, 1993).

What about surgical procedures? The standard book cited here is *Emergency War Surgery: NATO Handbook* (Pacific Publishing, 2011), which presupposes that even in battle conditions, there will be on-going links to modern medical technology and resources. Thus, there will be a supply of inhalational aesthetic gases such as such as isoflurane, sevoflurane, and desflurane, which will not occur in a long-term grid down situation. Previously used anaesthetics such as halothane, will ultimately suffer the same fate. There are of course opioids which can be used as anesthetic agents, so pain control is not hopeless. As well, maintaining surgical conditions to minimize the risk of sepsis, will be a challenge. But, the main difficulty will simply be that even the ordinary GP has limited surgical knowledge, and will only be good for relatively simple surgical procedures. The medical autodidact will probably be fine with simple things, such as wound suturing (stitches), extracting foreign objects from the skin, removal of ingrown toe nails/finger nails and activities like that. But, most surgery that is taken for granted now, such as precision eye surgery (e.g. pterygium excision with conjunctival autograph), will be beyond the capabilities of all but qualified surgeons. As for heart surgery, brain surgery, and transplants, forget it; if you need this in a long-term grid down scenario, kiss your ass goodbye.

It would be wise to get whatever surgery you have been putting off (e.g. say a carpal tunnel nerve release operation) done before it all falls apart. That goes for things like wisdom teeth, which usually cause problems at some point. You do not want to have to have someone try and pull these out, which will be like having your throat pulled out via your mouth. Getting all dental problems fixed now, practicing sound dental care, with a diet free of cane sugar, may be the wisest thing you can do in preparation for the end. Toothache will come back to haunt the remnant as a true curse of mankind. Don't be afraid of having any problematic teeth removed, for it is better that a dentist does it safely now, than Aunt Gertrude with her trusty, rusty pliers. Trust me; you do not need too many teeth to get by. And, go for extraction over root canal treatments for reasons detailed here.[175]

MANHOOD: GETTING LEAD BACK IN YOUR PENCIL

Apart from engaging in a massive education program about the survival aspects discussed in the last section, a person who intends to give it his best to survive the coming collapse needs the right mental attitude, philosophy and mind set. This is a difficult topic to approach since even though there are writers who have been through economic collapse situations, the sort of global cascading converging and compounding catastrophes discussed here are only now beginning to be experienced, and we have not seen anything yet.

Nevertheless, it is clear that survival will require a return to masculine values associated with warriorhood. Our present world is one where in the mad lands of the disUnited Kingdom, men who identify as women will be given a cervical smear test, even though they do not have a cervix![176] We live in a time where academic feminists make a good living from deriding all aspects of Traditionalism, especially masculinity, which is open game. This bs has caught on, as now the American Psychological Association has labelled traditional masculinity, the core values being strength, courage, mastery, and honor (per Jack Donovan), as "harmful."[177] Never mind that it is only masculine values embodied in the potential for violence, through police forces and the defense forces, which stand as a foundation for the protective structure of society, enabling thin-armed pencil necks to philosophize about the nonsense that they do.[178] At least one good

---

[175]   https://drjockers.com/root-canals/.

[176]   https://www.thesun.co.uk/news/5341098/men-who-identify-as-women-are-invited-for-cervical-smear-without-a-cervix/.

[177]   https://townhall.com/tipsheet/briannaheldt/2019/01/07/american-psychological-association-labels-traditional-masculinity-as-harmful-n2538637; https://www.apa.org/monitor/2019/01/ce-corner.aspx.

[178]   Jack Donovan, "Violence is Golden," http://www.jack-donovan.com/axis/2011/03/

thing coming from the collapse is that all of this bullshit will grind to a merciful halt, and the natural primal order will be restored, where natural selection will be king, as it should be.[179]

Apart from the political attack upon traditional masculinity, men face a wide-ranging environmental assault from endocrine-disruptors, estrogen-mimicking chemicals (xenoestrogens), and phytoestrogens found in "foods" such as soy,[180] caffeine, lignans (cereal brans and beans), coumestans (split peas, pinto beans, lima beans) and even beer (8-prenylnaringenin). These chemicals can have ill-health effects for both men and women, but for men there is the problem of lowering testosterone, something which is already crashing in the West, along with sperm quality.[181]

There are numerous steps that can be taken to minimize contacts with endocrine disrupting chemicals, but is not possible to eliminate them completely, and some damage is inevitable. What one can do is to minimize one's consumption of phytoestrogens in one's diet, and at least for men, move to a "testosterone-increasing" diet, as described in numerous books, but these are exceptionally good: Dr Martin Katahn, *The T-Factor Diet*, (W. W. Norton, New York, 2001), and Chad Howse (with Dr Stephen Anton), *The Man Diet*, (Chad House, 2018). This involves moving right away from a grain-based diet, as depicted in the food pyramid, and consuming high-quality protein from meats, carbs from anti-oxidant-rich vegetables and fruits, and correct levels of fats. The T-diet genre significantly overlaps with the paleo diet revolution, which has arisen in response to the obesity epidemic, and an array of other diet/lack of exercise epidemics, including diabetes, irritable bowel syndrome, and various food allergy reactions, often produced by the chemicals used in processing and flavoring foods. The paleo diet maintains that we should eat like a caveman, high protein, high fiber, and whatever

---

violence-is-golden/.

[179]   http://www.returnofkings.com/88671/4-reasons-why-collapse-will-be-the-best-thing-to-happen-for-men.

[180]   https://www.energeticnutrition.com/vitalzym/xeno_phyto_estrogens.html.

[181]   https://www.rooshv.com/the-decline-in-testosterone-is-destroying-the-basis-of-masculinity.

fats come with it, but with low carbohydrate: Dr Sarah Ballantyne, *Paleo Principles*, (Victory Belt Publishing, 2017). Refined sugar, and even fructose in excess, must be avoided as being worse than the plague: sweet poisons,[182] so even fruits, especially fruit juices, need to be carefully monitored, as fructose can worsen insulin levels.[183] Excessive sugar consumption can have a wide range of ill-health effects, being linked to breast cancer and metastasis in the lungs.[184] However, there needs to be balance here, as whole fruits do contain numerous anti-oxidants and other beneficial chemical compounds. Natural news.com, is a good source for examining the health benefits of fruits, such as, for example, the anti-aging effects of pterostilbene in blueberries.[185] Obviously, completely abandon the consumption of alcohol, all recreational drugs, cigarettes and other racial poisons.

A study of the isotopes of bones of deceased people from the 15th to 17th centuries in Northern Ostrobothnia, Finland, found that 70 percent of the diet of these Nordics, was fish such as Baltic herring.[186] Today, men in Iceland live longer than even the Japanese, thriving on a manly diet of mainly fish, with few vegetables and little grains.[187]

The paleo-T diet amounts to eating foods such as quality meats, especially Omega 3 rich oily fish, organic vegetables with a focus upon those high in anti-oxidants and anti-cancer nutrients (e.g. broccoli rather than potatoes), and the elimination of all grains. Little has been written about diet in a long-term collapse situation; the general assumption is that grains will continued to be grown

---

[182]   D. Gillespie, *Sweet Poison: Why Sugar Makes Us Fat*, (Penguin, New York, 2013).

[183]   https://articles.mercola.com/sites/articles/archive/2015/02/18/processed-fructose-obesity-diabetes.aspx; https://www.mercola.com/infographics/fructose-overload.htm.

[184]   https://articles.mercola.com/sites/articles/archive/2010/06/19/richard-johnson-interview-may-18-2010.aspx; https://www.naturalnews.com/2019-02-14-excessive-sugar-consumption-dramatically-increases-your-risk-of-cancer.html.

[185]   http://www.natural.news/2018-11-01-blueberries-contain-a-specific-substance-that-can-prolong-your-life.html.

[186]   M. Lahtinen and A-K. Salmi, "Mixed Livelihood Society in Iin Hamina – A Case Study of Medieval Diet in the Northern Ostrobothnia, Finland," *Environmental Archaelogy*, vol. 24, 2019, pp. 1-14.

[187]   https://www.naturalnewsblogs.com/eat-fish-cheat-death/.

after all stores have been used up. But, growing, say, wheat, assuming that one does not develop a gluten allergy over the long term, is not an efficient use of agricultural land, compared to growing vegetables, or even many nuts. In the long-term, survivors will return to a paleo style diet of meat and fish and vegetables, with some in-season fruits to finish a meal. There will be no metabolic diseases, obesity, diabetes, or the other horrors of life in the modern human battery hen coops. Life will be short and sweet, instead of long and bitter.

However, in a collapse situation, it may not be possible to store enough correct food to sustain one's family, given the vast quantities of food that people need. It should be possible for people having a Sustainable Autonomous Base (SAB), to preserve meats, fruits and vegetables in quantities to see them over most hard times; there are numerous books on Amazon on this, simply search for "prepping, canning and preserving," or just "canning," and "preserving." The Organic Prepper[188] is also a very good source of information. Despite this, it may be necessary to also stockpile dried foods such as beans and various grains, simply for long-term security, as these foods last a long time, and are calorie dense. Even if these are sub-optimal foods in ideal conditions, sub-optimal foods are better than starving.[189]

Before moving on from the diet issue, there are some foods with a long shelf life that should be in a survival stockpile. Protein powders used in bodybuilding and powerlifting present a concentrated source of protein, dead easy to prepare (mix with water), with a long shelf life, providing that the powders are stored out of direct sunlight in a cool dry place. Along with vitamin tablets, these items can keep one going in hard conditions. Olive oil is a neglected survival food too; an old Italian who survived World War II told me that he, as an old man then (the story dates back to the 1960s), had hidden some large bottles of olive oil, and that helped keep him alive during the food shortages. It is good for one's cardiovascular health as well.

---

188   https://www.theorganicprepper.com/.

189   https://www.theorganicprepper.com/how-to-feed-your-family-when-youre-flat -broke/.

Thus, getting one's diet in shape, right now, if the would-be survivor is carrying excess fat, is the first thing to do, even before physical training. One needs to rebuild oneself. This is especially important since obese people have on average lower brain volumes.[190] While it could be argued that people with smaller brains may be more susceptible to becoming obese, it is more likely given the obesity crisis, that brain shrinkage could be caused by obesity.[191] If that is not so, then there must be more dumb manure-makers out there than was previously thought.

There is also an ideological dimension to all of this, with weaker types of men, the "soy boys," often having testosterone levels of 85-year-old normal men. So, stay away from soy.[192]

Now, with your diet under control, what next?

PHYSICAL TRAINING:
SELF DEFENCE AND UNARMED COMBAT

There is no animal in the world so treacherous as man.
– Michel de Montaigne (1533-1592)[193]

Physical training is something that should be undertaken as soon as possible by the would-be collapse-survivalist. Physical training has as its sole aim, the desire to make people more wolf-like and not to remain in a lamb chop state. While it is important to lose weight, it is not necessary to reduce body fat to six pack levels. In fact, bodybuilding does not aim to primarily produce strength and athleticism, but to generate a "cut" profile. This obsession with a ripped look is dangerous from a collapse perspective. Most

---

[190]   https://www.livescience.com/64454-belly-fat-brain-shrinkage.html.

[191]   https://www.zerohedge.com/news/2018-10-05/shocking-new-studies-find-young-americans-are-overweight-unhealthy-suicidal.

[192]   https://www.dailywire.com/news/22906/buzzfeed-guys-test-their-testosterone-levels-amanda-prestigiacomo.

[193]   Michel De Montaigne, *Essays Book II*, Ch. 12 Essais (1595).

professional bodybuilders, and sadly power athletes, produce their phenomenally cut bodies by hard training, but also with chemical supplements including various cocktails of steroids and maybe growth hormone and who knows what else, to transform their bodies into something once described by, I believe, comedian Clive James, as a "condom full of walnuts." They consume enormous meals, throwing down a quantity of food which will not be available on a sustainable basis. Needless to say, the said chemical additives will not be around for long, and the large muscles will soon atrophy.

Strength training, done right, which is drug-free, is one of the best things a person can do to get ready for the hard times to come. There is the obvious physical dimension, which I will write about, but strength training combined with a practical self-defense unarmed combat system, can improve one's mental attitude, making a person tougher. I say this, not from research work from journals, but from training people and seeing personal transformations, even in my own sons and daughters. Pumping iron can lead to being more like iron than soggy pasta.[194]

Certainly, from a physical perspective, humans have been in decline for the entire duration of the modern world, and even with nutrition, have lacked the harshness of life that made people like the Vikings, Romans and Spartans, men to be feared.[195] For example, in the 4th century BCE, Greek author Xenophon wrote a matter-of-fact account of how Greek soldiers rowed an Athenian warship on a journey of 236 km from Byzantium to Heraclea in a day. This was an average of 7-8 knots over a 12-16-hour trip. Modern rowers can only manage 6 knots, and then only for an hour before conking out. Hence, if this account is accurate, and there is no reason for thinking that it is not, as it was reported in a matter-of-fact way, the average Greek soldier was fitter than a contemporary elite rower. These soldiers sustained themselves on a sub-optimal diet of barley mixed with olive oil, washed down with wine. Genetic changes may

---

[194] https://www.theguardian.com/commentisfree/2018/sep/27/do-you-boast-about-your-fitness-watch-out-youll-unavoidably-become-rightwing.

[195] P. McAllister, *Manthropology: The Science of Why the Modern Male in Not the Man He Used to Be*, (St. Martin's Press, New York, 2009).

or may not have occurred, but these are not significant enough to account for the differences, which are more likely due to the average premodern exercising intensely from an early age.[196]

The elements of fitness include muscular strength and related, bone density, cardiovascular fitness, and flexibility.[197] There are numerous modes of training nowadays in gyms/fitness centers, offering a plethora of different approaches to achieving fitness and the body beautiful. There is High Intensity Interval Training (HIIT), Pilates ("Contrology"), yoga, Zumba and many more. Screw them! For the apocalypse one needs hard-core barbarian training, that is totally dedicated to developing fighting fitness, attempting to be the men that the Spartans, Romans and Vikings were. So, ok, you need flexibility, and all the stretches you will ever need are covered in the iconic stretching book by Bob Anderson, *Stretching*, (Shelter Publications, Bolinas, 2010), now in its 30th year of publication.

For strength development, nothing beats old school classical weight training using predominately barbells, but spiced with dumbbells and maybe kettlebells. Basic compound exercises are performed, such as the squat, bench press, deadlift, power clean, bent over rows, calf raises, sit-ups with weights, and maybe for entertainment, biceps curls, but usually it is more efficient to train large muscle groups, so the biceps are better hit with heavy rows and chin-ups. These basic exercises are detailed in numerous places on the internet, but a good beginner's book is Mark Rippetoe, *Starting Strength Training: Basic Barbell Training*, (Aasgaard Company, Wichita Falls, 2017). Next up have a look at Bobby Maximus and Michael Easter, *Maximus Body*, (Rodale, 2018). This will give all of the weight training routines you will need to get moving, and get results.

I spoke of "barbarian" training and this approach is represented well by Steve Justa in *Rock Iron Steel: The Book of Strength*, (IronMind Enterprises, Nevada City, 1998) and *High Plains, Heavy Metal, Iron*

---

[196]   P. McAllister, "The Evolution of the Inadequate Modern Male," *Australasian Science*, May, 2011, pp. 19-21.

[197]   https://survivalblog.com/survival-fitness-health-part-1-jbh/.

Master, Super Strength Bible! (Strongerman Productions, 2012). This is but one example of the old school strongman approach to strength athleticism. A book with even more relevance to unarmed combat training is Bud Jeffries, *Super Strength and Endurance for Martial Arts*, (Strongerman Productions, Lakeland, 2005). The thesis of these books is that there are various types of strength, all of which interact in a holistic fashion, so a great variety of hard-core training methods is necessary to produce the well-rounded strength athlete, who can do more than simply bench press substantial weight. As Bud Jeffries says, the "formula" is "the combination of low reps, high reps, overloading movements, off balance movements, and conditioning." (p. 19)

Thus, to be considered in addition to conventional barbell, dumbbell and kettlebell lifts, are activities such as walking with heavy weights (classic farmer's walk), dragging or pushing weights, such as cars or trucks, barrel lifting, log lifting, sandbag lifting, rock lifting, log throwing/caber tossing (as in the Scottish Highland games), flipping large truck tires (if available), and numerous functional grip strength exercises. A book which goes into great detail about this type of old school training is Brook Kubik, *Dinosaur Training: Lost Secrets of Strength and Development*, (The Author, 2006), and for lifters over forty years, *Grey Hair and Black Iron: Secrets of Successful Strength Training for Older Lifters*, (Brooks Kubik Enterprises, 2010). Indeed, the later book is well worth reading even by younger lifters, as there is sensible material on a number of dangerous lifts that often result in injury. Old Time Strongman[198] relives the tradition and philosophy of the strongmen of the past such as Arthur Saxon (1878-1921) (training with abbreviated, intense drug-free section). For a taste of the wisdom of the past, consult, Arthur Saxon, *The Development of Physical Power* (1931).[199] Another site[200] tells you all you need to know to start developing powerful hands and grip strength, very useful for both labor and combat in the world to come.

[198]   https://www.oldtimestrongman.com/.

[199]   https://web.archive.org/web/20120529122448/http://sandowplus.co.uk/Competition/Saxon/DPP/dppintro.htm.

[200]   https://www.functionalhandstrength.com/

The next thing to consider is the enormously controversial topic of martial arts training, primarily unarmed combat. This is an area which I have been interested in for most of my life, and I have come to have heretical thoughts about this. Bruce Lee (1940-1973), expressed doubts about the efficiency of classical Asian and Western martial arts systems in the 1960s and early 1970s, essentially arguing that all such systems have limitations built into their framework, such as classical Wing Chun, with its preoccupation with close range fighting and largely straight attacks. He put all of this down in his posthumous book, *Tao of Jeet Kune Do*, (Black Belt, 2011).

Along the same lines are the objections made by John Perkins (et al.) in *Attack Proof*, (Human Kinetics, Champaign, 2009):

Whether by plan or accident, self-defense training on almost all levels has become inadequate, overstylized, unnatural, and in many cases, too sportive. To teach large numbers of people in a short time, instructors boil down defensive moves into simple regimented, robot-like techniques that bear no resemblance to actual fighting. Similarly, some originally authentic systems of fighting have developed into highly artistic and dance like art forms that are appropriate for demonstration purposes only. (p. xi)

James LaFond's book *The Fighting Edge: Using Your Martial Arts to Fight Better,*[201] is the best book I can recommend on exposing the mythology of the martial arts. Apart from the limitations of the "martial arts," there has been a much more practical tradition of unarmed combat coming from World War II systems, as seen in the works of people like William Fairbairn and Rex Applegate, *Kill or Get Killed*, (Paladin Press, Boulder, Colorado, 1976). This has in turn led to many military hand-to-hand combat systems being marketed. In general, this is a vast improvement upon classical Asiatic approaches. However, as has been pointed out in the article, "The Myth of Military Hand-to-Hand Combat Systems,"[202] beyond the

---

[201] https://www.jameslafond.com/?f=store&id=48.

[202] http://www.wimsblog.com/2013/03/the-myth-of-military-hand-to-hand -combat-systems/.

sound fundamentals, the special combat units in the military do not have the time to develop high skill levels in unarmed combat, since they have other teaching priorities for soldiers who will be fighting with weapons. Sure, bullets run out, and this has often happened, but it is rare for soldiers in any era to go immediately for unarmed fighting, unless there is no weapon available. Usually a fighting knife would be resorted to in a last pitch effort. As Sgt Rory Miller has said: "Possibly the most overlooked aspect of power generation in the martial arts is one of the most effective: use a tool. I will take a hickory baseball bat over the hardest fist on Okinawa."[203]

Even with the rise of Mixed Martial Arts, which has shown the limits of the standard Asian systems, there are still many stories of superb fighters, capable of beating down multiple attackers, still being defeated, or worse, killed by punks armed with weapons, such as iron bars and knives.[204] And, regarding the present fad of ground fighting, while this is unquestionable effective in the octagon, it was never a good idea in the street, or bar, with sharp objects around, such as broken glass, and the knife of your opponent's buddy.[205]

In conclusion, the basic philosophy of the martial arts and even more practical unarmed combat training is that you will be weaponless, or lose your weapon, or won't have a spare, or any number of other hypothetical scenarios. But, while there are merits in having these unarmed skills, the point remains that the enormous amount of time devoted to become proficient in these unarmed skills, does not cash out in real self-defense benefits compared to weapons training. Perhaps that is why the Asian systems, which take years, if not decades to master, bullshit on about spiritual values and all the rest of it. While one can lose a weapon, better skilled fighters simply do not, have backup and backup to the backup weapons, and other plans. People actually did this in the past, and survived, with only basic unarmed skills, if that, perhaps nothing more than a

[203]   R. Miller, *Mediations on Violence*, (YMAA Publication Center, Boston, 2008), p. 138.

[204]   https://www.dailymail.co.uk/video/news/video-1833704/Video-Final-moments-hero-bouncer-fought-gatecrashing-mob.html.

[205]   https://guidedchaos.kartra.com/page/SteinerMythOfGrappling.

strong arm and the will to rip out an eye or crush a throat. Therefore, the sort of intense time-consuming effort required for unarmed self-defense training, from an apocalypse perspective, is best channeled into weapons training, firearms first, then melee weapons. If there is spare time, unarmed combat definitely should be studied, but it is not a chief priority. Developing survivalist weapons savvy is much more important. Sure, MMA fighters are touch guys in the cage going up against opponents playing by the same rules, but in the Great Collapse one needs to throw away the rule book.

There are many good books in the urban survival genre dealing with primarily surviving contemporary urban violence. James LaFond has produced a number of books derived from surviving in one of America' most violent cities, Baltimore, and along with his combative books, he has case studies of predatory violence, including, *When You're Food*, *Waking Up in Indian Country*, and *Thriving in Bad Places*, which take "situational awareness" to a whole new level.[206] Also well worth reading is Tim Larkin, *When Violence is the Answer*,[207]("Your brain is your deadliest weapon"); Marc MacYoung and Jenna Meek, *What You Don't Know Can Kill You* (2018),[208] and Varg Freeborn, *Violence of Mind* (2018).[209] If necessary, from there the once novice reader can seek out material on his own to digest. It is far better to learn lessons from the stab and gunshot wounds of others, then to learn from firsthand experience, where, as luck will have it, you die.

The main lesson to take on board for the collapse, is that the world ceases to be safe in the ways people have taken for granted, for the world will return to a pre-modern state of "constant battles."[210] In the die off period following the collapse, the adrenaline level will be higher, but even after the bulk of fetid humanity has

---

[206]   See: http://jameslafond.blogspot.com/p/ssurvival.html.

[207]   T. Larkin, *When Violence is the Answer*, (Little Brown and Company, New York, 2017).

[208]   M. MacYoung and J. Meek, *What You Don't Know Can Kill You*, (Carry On Publishing, Monument, 2018).

[209]   V. Freeborn, *Violence of Mind*, (The Author, 2018).

[210]   S. LeBlanc, *Constant Battles: Why We Fight*, (St. Martin's Press, New York, 2003).

become food for rats, the remnant will still face danger from other survivors, who will continue to be threats so long as resources are scarce. Thus, the individual would need to live in a state of relaxed vigilance, being aware of one's environment and what is around him; typically described as condition yellow in Jeff Cooper's color code of situational awareness.[211] Condition white is a state of sleepwalking unawareness, a mechanical doing that would leave it up to luck if one survived. Condition orange, next beyond condition yellow, is a focused awareness on something that could become a threat, and condition red, is what it sounds like, red alert, battle stations. Many US police go through this color code awareness (yellow, orange, red, yellow …) each day, and if they can do it, so can you.

A book dealing with the brutality of collapse and the war in the Balkans, which does deal with the aspects of hyper-violence, is Selco Begovic, *The Dark Secrets of SHTF Survival*.[212] This book is particularly good for its hard, gritty realism about the level of degradation that people can fall to, and it refutes liberal-humanist myths of the milk of human kindness, and that the good of human nature ultimately shines through dark situations (my interpretation). There is of course a limit from what one can obtain from books, but at least such literature can get one's thinking clear about the topic, which is a good first step.

Psychologist Richard Logan, writing in the foreword to Les Stroud's *Will to Live*, says that he agrees with Stroud's central thesis that "sometimes the only explanation for why some people survive a hell beyond hell is their steely will to live. Survival often goes so far beyond the capacity of psychologists to explain that there is simply nothing left to say."[213] The will to survive is something which either burns within one, or does not, much like other virtues such as courage. Probably it cannot be created *ex nihilo*, but physical training may enhance the flame if it exists.

[211]   https://cdn.ymaws.com/www.caceo.us/resource/resmgr/imported/documents/as14/Coopers-System-for-Awareness.pdf.

[212]   S. Begovic, *The Dark Secrets of SHTF Survival: The Brutal Truth about Violence, Death, and Mayhem that You Must Know to Survive*, (Daisy Luther Media, 2018).

[213]   L. Stroud, *Will to Live: Dispatches from the Edge of Survival*, (Harper, New York, 2011).

WEAPONS

Good fighting came before good writing.

– John Marston (1617)[214]

Firearms, and proficiency in their use, should dominate any present survivalist concern with preparing for the coming collapse. However, the future of private firearms ownership is grim, even for us Americans. Other jurisdictions have severe firearm restrictions, and the "progressive" elites in America are conducting a multi-faceted program to ultimately lead to US gun confiscation, from ammunition sales restrictions, to the criminalization of private gun sales, to outright Australia-1996-style confiscation, perhaps using the armed forces to brutalize civilians. [215] The later scenario hypothesizes that the Democrat president will use the *National Defense Resources Preparedness Executive Order 13603*, issued by President Obama on March 16, 2012, to declare a "national emergency" over gun violence, with the use of emergency powers to conduct a nationwide gun confiscation program. There are doubts on some sites whether this Order does permit this,[216] but if it does not, surely the "correct" Executive Order will be produced by said Democrat president.[217]

Facing the coming apocalypse described above will be worse than horrendous without firearms, and even now in most of Europe, where the bar is set very high to own any firearm, gun ownership is surging.[218] Places such as Venezuela, where the sheeple foolishly allowed gun confiscation, now face threats in both urban and rural

---

[214]   John Marston, *The Mountebank's Masque* (Paradox XV, 1617).

[215]   https://mises.org/wire/bipartisan-support-new-federal-gun-controls-red-flag; http://www.shtfplan.com/headline-news/incoming-communists-ready-a-bill-that-will-criminalize-private-gun-sales_12182018; https://www.naturalnews.com/2019-01-27-oregon-democrats-push-new-bill-to-outlaw-self-defense-violent-crime.html.

[216]   https://en.wikipedia.org/wiki/Executive_Order_13603.

[217]   E. Kaufmann, *Whiteshift: Populism, Immigration and the Future of White Minorities,* (Allen Lane, London, 2018).

[218]   https://www.wsj.com/articles/gun-use-surges-in-europe-where-firearms-are-rare-11546857000.

areas from gangs, often armed with illegal guns, who are preying upon the vulnerable population, with little protection from the corrupt police force.[219]

In jurisdictions where private firearms ownership is severely restricted, it may still be possible to attend rifle, shotgun and pistol clubs for novices to get some basic competency with these weapons. It would be important to move beyond merely shooting at paper targets, whether using a rifle, shotgun or pistol, and engage in force-on-force realistic training, to safely simulate real world fire fights, and to attempt to reduce the problem of misses under stress.[220] NYPD statistics for 1994-2000 have a combat situation hit percentage at zero to two yards at 38 percent, with 62 percent misses, due to the effects of adrenaline and fear in general.[221] There is a wealth of information on the topic of realistic firearms training on YouTube, and internet sites.[222] A good book to get one's feet moving on close quarter combat is Shawn Williams, *Armageddon CQB: A Close Quarter Battle Primer for the Apocalypse*, (Outskirts Press, 2018).

If all of the above training is impossible, as seems to be the case in parts of East Asia, there are still internet resources and numerous books to consult, that, while far from perfect, may be the best possible under the circumstances for learning. Some to read include, Don Mann, *The Modern Day Gunslinger*, (Skyhorse Publishing, New York, 2010); Wayne Van Zwoll, *Mastering the Art of Long-Range Shooting*, (Gun Digest Books, Iola, 2013); Joe Nobody, *The Home Schooled Shootist: Training to Fight with a Carbine*, (Kemah Bay Marketing, Kemah, 2011); Bill Jordan, *No Second Place Winner*, (Police Bookshelf, Concord, 1989) and Jeff Cooper, *The Art of the*

[219] https://www.theorganicprepper.com/defending-homestead-venezuela/.

[220] M. Greenman, *The Zombie Shooting Guide: Survival Training for the Worst-Case Scenario*, (Ooda Media Group, Los Angeles, 2013).

[221] R. Miller, *Meditations on Violence: A Comparison of Martial Arts Training and Real World Violence*, (YMAA Publication Center, Boston, 2008), p. 58.

[222] Of interest: https://survivalblog.gunfighters-guide-lessons-learned-hard-way-part-1-grumpy-gunfighter, and https://survivalblog.com/gunfighters-guide-lessons-learned-hard-way-part-2-grumpy-gunfighter/; https://survivalblog.com/gunfighters-guide-lessons-learned-hard-way-part-3-grumpy-gunfighter/.

Rifle, (Paladin Press, Boulder, Colorado, 1997). That would give a beginner a basic introduction, enabling them to then explore other books via Amazon.com, and numerous gun and self-defense sites. There is still a vast amount of excellent information on YouTube on gun-related training, and it should be watched while it is still there. I agree with gun guru the late Elmer Keith (1899-1984), the "father of big bore hand gunning," when he summed up the firearms/self-defense training issue thus: "Anyone can be taught in a couple of hours to aim or point a shotgun, heavy caliber six gun or the .45 Auto at the middle of a criminal, and score decisive hits at the first shot. Aim or point any of these weapons at his middle and shoot as long as he is on his feet. The remaining trouble will then be up to the undertaker."[223] It is ballistics, but not quite rocket science.

Numerous books written since the 1970s have proposed that at a minimum, one needs for self-defense, a military-style semi-auto rifle (e.g. AR-15), a combat pistol, generally a semi-auto rather than a revolver, and a 12-gauge shotgun. This is the position of Dr Bruce Clayton,[224] Mel Tappan in his iconic *Survival Guns*,[225] and Fred Rexer,[226] to name but a few. That is a basic ideal battery for one person, and primarily for self-defense against human attackers; Tappan and others also believed that numerous "working" guns are also needed, to deal with hunting small game (.22 LR), or defense against large predators such as bears.[227] There are debates in most gun magazines, or gun comics if you are cynical, about the merits of various semi-auto pistols for self defense, both brand and caliber, perhaps generated by the intrinsic limitation of most pistols compared to heavy caliber rifles and shotguns, to have the

---

[223]   E. Keith, "The Best Home Defense Gun," *Guns & Ammo Guide to Guns for Home Defense*, 1975, pp. 42-44, at p. 44.

[224]   B. D. Clayton, *Life After Doomsday: A Survivalist Guide to Nuclear War and Other Major Disasters*, (Paladin Press, Boulder, Colorado, 1992).

[225]   M. Tappan, *Survival Guns*, (Paladin Press, Boulder, Colorado, 2009).

[226]   F. Rexer, *USA: The Urban Survival Arsenal*, (Delta Press, 1980).

[227]   https://survivalblog.com/weapons-systems-approach-firearms-training/; https://survivalblog.com/well-balanced-gun-collection/; http://www.survivopedia.com/top-6-survival-rifles/.

mythical "one shot stopping power."[228] For example, many survival savvy people swear by the reliability of the Glock family of semi-auto pistols, such as the Glock 17, with its 17, or even 33 round extended magazines. But, some have abandoned their Glocks and moved to the SIG Sauer camp, such as the SIG P320, allegedly being better for concealed carry.[229] These sorts of debates are somewhat "academic," since pistols are a backup in a grid down situation, where the rifle and shotgun are kings, and most popular commercial brands of defensive pistols are completely adequate for survival self-defense, even if such pistols would not pass an army small arms torture test. One is lucky to have any pistol at all.

Most of the survival gun literature is American, and assumes that the Second Amendment will survive. Less dramatically, there is an assumption that semi-auto rifles and pistols will still be legal in civilian hands. Yet, if America goes the way of, say, Australian, New Zealand, or even Britain, then Americans will have to make do with whatever repeating weapons are available. As noted by James Ballou in *Arming for the Apocalypse*,[230] survivalists would need to suffice with various hunting rifles, including bolt action and pump action repeaters, and the iconic lever actions from Marlin and Henry. There will be no room here for a "spray and pray" approach that could come from the abundance of riches of the semi auto, and marksmanship will have to rule. Perhaps this is not all bad, since attacking ferals may "surrender" their possibly semi-auto firearms, so that these guns could be seen as step ladders. But if not, these non-auto guns are highly reliable, not dependent upon gas and other auto-ejection systems, and fire generally more powerful rounds, or at least rounds where there is less controversy about stopping power. These guns could serve the role of a "scout rifle," as conceived by Jeff

[228]   P. A. Kasker, "Terrible Reality," *American Survival Guide*, February, 1993, pp. 16-17; W. U. Spitu (et al.), "Physical Activity Until Collapse Following Fatal Injury by Firearms and Sharp Pointed Weapons," *Journal of Forensic Science*, vol. 6, 1961, pp. 290-300.

[229]   https://www.naturalnews.com/2018-11-04-why-i-switched-from-glock-to-sig-for-concealed-carry.html.

[230]   J. Ballou, *Arming for the Apocalypse*, (Paladin Press, Boulder, Colorado, 2012).

Cooper,[231] a general-purpose repeater rifle for both hunting and self-defense, powerful enough to put down living targets of up to 1,000 lbs/ 454 kg. Ruger, along with other firearms firms, has marketed an excellent version of the scout concept, the Ruger Gunsite Scout, in .308 caliber. Nevertheless, ignoring dangerous game such as bears, for self-defense against human foes, even lever action rifles from Marlin, Henry and other firms producing cowboy action guns, such as Uberti and Cimerron Firearms, firing rounds in .357 Mag., .44 Mag., .45 Long Colt, will be adequate. The Browning BLR, has a box magazine, and comes in a variety of calibers with spitzer bullets (i.e. pointy nosed), such as in .223 and .308. However, Hornady's LEVERevolution rounds with special tips, allows the firing of a spitzer bullet in a rifle with a tubular magazine. The Remington 7600 Police Patrol pump action rifle comes in .308 and .223 calibers, and the .223 version takes AR-15 magazines, making it a semi-politically correct "assault rifle."

Those who are concerned about gun banning have generally proposed that survivalists have a crossbow, which is far easier for gunmen to use than a compound, long or recurve bow.[232] The bad news is that in some countries where gun bans have occurred, attention has then shifted to crossbows. Australia is an example, which shows that gun control leads to total weapon control. Indeed, in the state of Victoria, Australia, even swords are banned, and all Australian states heavily regulate classical Asian martial arts weapons, ironically the same sorts of weapons that the Okinawans used when disarmed by the Japanese in the 15th century. In any case, while bows have been used for longer distance shots, for self-defense fast shots are required in many cases. The excellent video, "A New Level of Archery,"[233] featuring Danish archer Lars Andersen, shows how a return to classic shooting, perhaps as practiced by the rapid-fire shooting of the archers of Genghis Khan[234] and the Comanches,

231   http://jeffcoopersscoutrifles.blogspot.com/.
232   https://urbansurvivalsite.com/survival-bows/.
233   https://www.youtube.com/watch?v=BEG-ly9tQGk.
234   https://www.youtube.com/watch?v=6lf9q6OQseo.

can lead to the ability to fire 10 arrows in 4.9 seconds, and 3 arrows in 0.6 seconds. Critics have said that these techniques are only valid for close-range rapid fire, but even so, many of our self-defense concerns are in this basic pistol range. Mastering these skills, even in part, would give one an advantage in a showdown against multiple attackers, perhaps in some out-of-doors bug out situation where firearms were not involved. Certainly, if one could approach that rate of fire, the bow is viable as a self-defense projectile weapon in gun-banned jurisdictions. As far as I can ascertain, recurve bows have not yet been restricted in jurisdictions where crossbows are now highly regulated.

If one does not have the time to develop such skill, in a firearm-banned society that has not yet banned crossbows, perhaps the recently released fast and accurate Cheap Shot 130 crossbow by Cold Steel[235] may be considered. This delivers shots capable of killing boar and deer within reasonable distances, with a rapid fire of a demonstrated 6 shots in 39 seconds, something which could be bettered. The "Buzz Saw" broad head can slice clean through a 200 lb pig carcass, so long pig would be no problem. It is probably not possible to fire arrows at Lars Andersen's rate, but as a range weapon against non-armored attackers in a rural environment/outdoors, with no firearms, it would be devastating. The Cobra System RX Adder, is a tactical crossbow being a modernized version of the "chu-ko-nu" weapon, having a top-level magazine with five (or six) Ek carbon bolts, which is worth a look.[236]

As far as range/projectile weapons go, if archery falls, following guns, then thrown weapons are a fall back, ranging from homemade thrown spears, such as sharpened sticks if necessary (native peoples did well even with throwing sticks), to a variety of factory made weapons such as Cold Steel's Torpedo Thrower 80 TOR, a piece of rolled carbon steel, with points at both ends.[237] A little expensive for

[235] https://www.youtube.com/watch?v=-CNaUbG54R8; http://insidearchery.com/coldsteel-cheapshot130/.

[236] https://www.youtube.com/watch?v=P2WhUu15B5E; https://hiconsumption.com/2019/03/ek-archery-cobra-rx-adder-tactical-repeating-crossbow/.

[237] https://www.youtube.com/watch?v=5qKP9TXHwrE.

the apocalypse in my opinion, where many such weapons will be needed, with most getting "lost." An alternative would be to make one's own, using thick steel rod, reo rod, sharpened. But, many things can be thrown, and simple rocks have been used since man crawled out of the primal slime, dripping filth, to kill his neighbor and steal his woman, food, and stuff. One can also cheaply make throwing weapons being essentially arrow-like darts with weighted tips, that will always hit the business end first, and one can hold a number of these in hand.[238] Just for the apocalypse, of course, not now!

As far as melee weapons go in an apocalypse scenario, most of the discussion of their use derives from zombie apocalypse literature and film. I am somewhat tired of the zombie genre, which like zombies does not die a natural death. In any case, assuming that killing zombies is a metaphor for fighting post-apocalyptic feral scum, we can do much better in the real world than using baseball bats and crowbars, and a variety of sporting goods, or even kitchen utensils, as some of the zombie books suggest. Yes, many people have been killed by baseball bats, but the sort of bat picked up at the sporting goods store today, really cannot take much by way of zombie head-beating, and the relatively lightweight maple bats, that have replaced ash, and before that hickory, often break on the field from hitting mere baseballs, rather than skulls.[239] Aluminium bats, more expensive, may be more durable, but can still warp, and in torture tests, like demolition events, eventually break, often surprisingly quickly.[240] The synthetic bats produced by Cold Steel (e.g. Brooklyn Crusher and Smasher) are very good, and have reviewed well. However, mattock handles, especially synthetic ones are also good, and cheaper. Indeed, axe handles are cheap and arguably better weapons than baseball bats, and are designed for taking impact.

As for crowbars, these are highly durable, having primary uses as a tool other than as a weapon. But, if they have reach, they will be somewhat weighty and unwieldy. These may be fine for dispatching

---

[238]   https://www.youtube.com/watch?v=rdYjYnEOR_w.

[239]   https://www.livescience.com/2699-science-breaking-baseball-bats.html.

[240]   https://www.youtube.com/watch?v=pv4D8AjUT1w.

the odd zombie, but will be slow against human opponents who may have superior, faster weapons. As for anything found in the kitchen, to make utensils a primary weapon, rather than a weapon of the moment (while cooking), is asking to die, probably unfed too.

The better zombie apocalypse books see the most appropriate anti-zombie melee weapons as being precisely the medieval weapons used against humans in the past, various polearms such as spears, swords, axes, war hammers, and large knives, as last-ditch weapons.[241] There are, of course, many different types of melee weapons, including flexible weapons such as three section staffs, and various whips, but for ordinary people rather than specialized martial artists, it is best to keep to the tried and proven more conventional weapons, rather than something like a three section staff, where one is more likely to injure oneself than one's opponent.

For most people, melee weapon choice will involve bladed weapons rather than rigid body impact weapons, such as various sticks and staffs. It was noted by the European masters, such as Joseph Swetham in *The Schoole of the Noble and Worthy Science of Defence*, (London, 1617) and earlier by George Silver, *Bref Instructions Ypon My Pradoxes of Defence* (London, 1599), with all other things being equal, a good quarterstaff man would defeat an unarmored, non-shield carrying, blade-wielding, opponent of "equal" ability. The reach and lightness of the staff gave a speed advantage that would overcome the merits of say a blade, or even a heavier pole weapon such as a poleaxe. Still, today, most people are not going to acquire this skill level[242] given all of the demands facing us, and the collapse will only intensify this. However, pole training can be an important aid to power training for whatever weapon one uses, especially if large thick steel pipes are used as

---

[241]   R. Ma, *The Zombie Combat Manual: A Guide to Fighting the Living Dead*, (Berkley Books, New York, 2010); A. Alasdair, *Weapons and Warfare in the Zombie Apocalypse*, (The Author, 2012).

[242]   See though for those wanting to acquire master of the fighting staff: D. C. McLemore, *The Fighting Staff*, (Paladin Press, Boulder, Colorado, 2009), and J. Varady, *The Art and Science of Staff fighting: A Complete Instructional Guide*, (YMAA Publication Center, Wolfeboro, 2016).

leverage bars, to increase upper body and forearm strength. Even in a collapse situation, without barbells and conventional weights, one could still do an upper body workout using such bars, thick scrap iron, or branches from trees or logs. Heavy clubs were used by the Indian wrestlers to increase strength, and the same can be done in a post-apocalyptic environment, using found material. A log can have a handle carved into it to enable safe grip, and this can be slung as one would if using it as a weapon, making sure not to clout yourself in the head. This will in turn increase the destructive power of blows delivered from bladed weapons, even if one does not use sticks as a primary weapon.

Another argument in favor of bladed weapons, if they are available, is that for the average person, it may not be possible to put down some lunatic with one hit from a stick before he is on to you. Remember the Tueller drill, that for someone with a holstered gun, there is 21 feet (6.4 meter) danger zone facing an opponent with a blade, who may bridge the gap before he can be decisively stopped by drawing and firing a holstered pistol.[243] This will also apply to the use of stick weapons. Basically, the knife wielding opponent could be onto one, if the first attack fails. Of course, unlike the pistol defense, there is always footwork, blocking and counter-attacks, but these could in turn come unstuck. For the moderately trained person who is not going to have the combat strategic sense of a decades-trained martial artists, in a post-apocalyptic scenario, there will need to be use made of weapons capable of delivering to an unarmored person a one shot stop.[244] While that concept may be problematic for firearms, blades can deliver at closer range devastating attacks, including complete decapitation, and cutting off limbs, and slicing a person in halves, head down to crotch. There are numerous YouTube videos from Zombie Go Boom.com, where synthetic ballistic heads are easily cut in half by bladed weapons, such as swords, and more dramatically, Cold Steel usually tests its bladed weapons by cutting demonstrations using animal carcasses, where even the lighter machetes (e.g. the Latin machete) and knives, go well in slicing

243   D. Tueller, "How Close is Too Close?" *S.W.A.T. Magazine*, March 1983.

244   https://www.buckeyefirearms.org/alternate-look-handgun-stopping-power.

through flesh. Just as enjoyable, for those with a Nordic/Germanic/ Viking taste, are the YouTube videos by Thegn Thrand, where medieval weapons are put through their paces,[245] with many tests of reasonably priced swords from Australia's Medieval Shoppe,[246] a most reasonably priced business that mail deliver their goodies to us here in America.

Cutting to the chase, my recommendation for the would-be survivor is to obtain a sword that one likes, and a spare, or if swords are illegal in one's jurisdiction, perhaps fighting axes, such as the Viking hand axe. Longer polearm weapons, such as spears, certainly have their place, the spear being man's primary melee weapon defense against wild animals. Some should be retained at one's survival retreat, perhaps duct taping a fighting knife such as a Bowie to a stout pole. Of course, commercial spears can be purchased, but these are getting pricey for what one gets, but there is likely to be some medieval re-enactment society in your town, with a blacksmith who can bang out spear heads that you then attach to a pole. The quality of workmanship, as the Vikings knew, need not be as precise as that required for swords. Likewise, for axe heads, which may be obtained from the same source.

With swords being banned in various jurisdictions, fighting axes such as the Viking hand axe are an option for melee weapons. So, how good would the Viking axe be as a post-apocalyptic weapon? John Clements sees the fighting axe as having disadvantages relative to the sword, as the long axe allegedly lacks versatility and proper defense and thus could be defeated by a swordsman with a shield.[247] He is probably referring to the two-handed Dane axe here rather than the smaller Viking hand axes, which are even more versatile than swords, capable of being grasped close to the blade for close-range fighting, as well as being able to be thrown, something not usually done with a sword. These light axes can chop better than

---

[245]   https://www.youtube.com/user/ThegnThrand?pbjreload=10.

[246]   https://medievalshoppe.com.au.

[247]   J. Clements, *Medieval Swordsmanship: Illustrated Methods and Techniques*, (Paladin Press, Boulder, Colorado, 1988), p. 177; J. Kim Siddorn, *Viking Weapons and Warfare*, (Tempus Publishing, Gloucestershire, 2003).

most swords, if long enough, and are not slow in recovery, as the larger Dane axe, or even a wood chopping axe would be. These small axes feature in the Bayeux Tapestry and other depictions, so the Vikings, Normans and Germanic warriors must have used them as weapons of war. Further, these sorts of axes are still pretty much under the politically correct radar weapons, and can be purchased for much less than any decent sword. A family could arm up with them, while obtaining good swords, today, is still an expensive venture. Expense and availability were some of the main reasons for Vikings preferring axes to swords, and for us poor people today, the same argument holds. For axe training, including fighting tomahawks, there is not much good material on YouTube, swords overwhelmingly dominating. For books, D. C. McLemore, *The Fighting Tomahawk: An Illustrated Guide to Using the Tomahawk and Long Knife as Weapons*, (Paladin Press, Boulder, Colorado, 2004), and *The Fighting Tomahawk* Volume II, (Paladin Press, Boulder, Colorado, 2010), can serve as introductory texts. The fighting axe techniques are much easier to learn than sword fighting, and people who have used woodwork axes already have a head start.

Swords have a special place in the dark hearts of doomsters as melee weapons. Indeed, as John Clements has observed, swords are a quasi-mystical weapon of high symbolic significance, often having names, and featuring in epic sagas of light versus darkness and other English lit stuff that we forgot from school.[248] Apart from that, swords deliver truly devastating attacks upon unarmored human flesh, and as has been said above, while there could be problems for the non-expert putting down some drugged lunatic, even with a handgun, a good sword has no such erectile deficiency problems. The Moro suicide attackers, the Juramentados, during the 1899-1913 Moro Rebellion in the Philippines, would often soak up numerous slugs from the US .38 caliber service revolvers of the time, then kill the US soldier with a short sword. Legend has it that this led to the use of the Colt .45 Model 1911 semi auto pistol, but this service side

---

[248] J. Clement, "Echo of Steel," in S. Shackleford, *Spirit of the Sword: A Celebration of Artistry and Craftsmanship*, (Krause Publications, Iola, 2010), pp. 32-51; H. Withers, *The Illustrated Encyclopedia of Swords and Sabres*, (Lorenz Books, Leicestershire, 2012).

arm arrived after the Moro Rebellion. Instead, the Colt .45 Model 1902 and DA Model 1909 revolvers were used, which were better than the .38 revolver, but not perfect against drugged-up crazies.[249] Winchester repeating shotguns fared better. However, the use of a sword, longer than the Moro's one, be it katana, Viking sword, or other European and Asian blades (such as the fearsome Chinese war sword or *Dadao*[250]), would decisively end the confrontation, since no-one fights on after being decapitated.

It is also interesting to note that today the Philippines' marine corps and Special Forces issue soldiers with short swords such as the bolo and ginunting, with blades up to 51 cms (20 inches). The weapons are useful for fighting in thick jungles in Mindanao, where very close-range fighting can occur. These weapons are quiet and effective in the vegetation, which offers a parallel situation for civilians in a domestic home invasion scenario, where the actual house presents the cluttered environment, that may not be conducive to using a rifle (even a carbine), or shotgun. While swords have long been abandoned from most other armies, in favor of the use of bayonets, that does not necessarily apply to civilian survival use, where the issue of weight and interference with other gear, may or may not be applicable.

The sword which a survivalist chooses for a multigenerational conflict is really a matter of personal style and taste. In general, one would be wise to have a sword which was excellent in stabbing and slashing/chopping, but most swords are good enough here; even the Chinese war sword can stab.[251] The real issue in my opinion is not the futile question of "what is the best sword," but rather, what type of sword best fits into your melee combat strategy. For example, there is no question that the huge two-handed "great sword"[252] or the *Zweihänder*, used by the German Landsknecht mercenaries in

[249] https://www.manilatimes.net/juramentados-and-the-development-of-the-colt-45-caliber-model-1911/107609/.

[250] https://www.youtube.com/watch?v=8PQiaurIiDM.

[251] https://www.youtube.com/watch?v=8PQiaurIiDM.

[252] https://www.youtube.com/watch?v=_hfLZozBVpM.

the early 16th century to supposedly chop the heads off of pikes, would be effective in some outdoor situations, especially in a rural environment. But, in general, such a weapon is going to be clumsy in most applications, and a more practical sword will be needed. I believe that whatever sword one likes, and up to a point, there will be a strong element of personal taste and aesthetics in sword choice, the weapon should be capable of effective one-handed use. This is necessary for use of a shield, even one banged up from marine plywood, to defend against the use of throwing weapons, such as spikes, weighted arrows, throwing sticks and even rocks. What point is there having a huge blade wielded in two hands, if one gets one's skull smashed in by a rock, in a David vs Goliath death match?

It should be possible to train to be able to use iconic two-handed swords, such as the katana, with one hand. Japanese swordmaster Miyamoto Musashi in *A Book of Five Rings*[253] had a method of two-sword fighting, *Niten Ichi-ryū* (二天一流), "the school of the strategy of two heavens as one." This involved use of the katana and the shorter wakizashi, indicating that the katana could be effectively wielded in one hand, and would achievable by strength training, and using training "swords" such as iron bars, much heavier than the actual sword. It is not too difficult.

Obviously, if one opts for use of a sword, a fine weapon is needed, and here is the age-old problem with swords, even for today: expense. A cheap made-in-China knockoff katana, is probably going to be unreliable, although I did buy a Damascus steel katana which cuts very well, but who knows when TSHTF. Cheaper swords bought from on-line shopping sites may be suspect, for your life could one day depend upon their reliability. Thus, if finance is going to be an issue, as I expect that it will for many people today, struggling just to get by, the alternative could be a combat machete. Cold Steel makes numerous permutations of a basic pattern of 1055 carbon steel with baked-on anti-rust matt finish, and 2.8 mm thick blades. There are classic working machetes such as the Latin, but others are along the cheap weapons line, such as the gladius, tactical katana, Chinese

---

[253]   Miyamoto Musashi, *A Book of Five Rings*, (Overlook Press, Woodstock, 1974).

war sword, barong, kopis, various Bowies, and kukris, some with neat hand guards, such as the Royal kukri. One should go to their website, have a look at what strikes you as interesting, then seek out a store that sells them, try it in your hand, and purchase if satisfied. Obviously, there are numerous other brands of machetes to choose from, but Cold Steel comes first to mind as giving good bang for the buck. Nevertheless, for workhorses that can double as weapons, the Latin and bolo machetes produced by the Brazilian company Tramontina, may be even more value for money.[254] MTech's Tactical Combat Machete, with a 25-inch, 4 mm thick blade, and Kershaw's Camp 18, would also be excellent cost-effective choices for machetes that are really short swords, so there is no sharp dividing line between machetes and swords.[255]

A good book to get started on sword/machete training is Fred Hutchinson, *The Modern Swordsman: Realistic Training for Serious Self-Defense*, (Paladin Press, Boulder, Colorado, 1998). This book does not single out any particular type of sword to be used for self-defense, but rather offers general training strategies based on practical methods used in warrior cultures of the past, such as ancient Rome. This involves using a surrogate sword for training much heavier than the battle sword (e.g. steel rod), strikes on a training post or hanging tire or heavy bag, safe sparring with mock weapons and protective gear with a partner(s), and a devotion to physical fitness and all-round athleticism. Weight training for the wrists and forearms should be conducted. Speed training, such as striking a ball on a string, should be done.

Well worth looking at while it still remains on YouTube is Cold Steel's take on machete fighting, which gives useful insights for blade fighting in general.[256] I do not have a high opinion of the Asian martial arts, and so as far as weapons training goes, one needs to get skills taught at a reasonable pace, rather than being made into

---

[254] https://www.machetespecialists.com.

[255] https://www.survivopedia.com/how-to-defend-against-a-machete-attack/.

[256] https://www.youtube.com/watch?v=rzbPdJEZw6o; https://www.youtube.com/watch?v=EhUzkdntyqs.

a cash cow for "the Sifu." Perhaps, if HEMA (Historical European Martial Arts) is in your town, it would be worth checking out. There could even be medieval sword fighting clubs at the local university. Both would supply training partners for free sparring, which is the key thing to getting good in any combat weapons-based activity; you have to fight a lot.

Finally, the knife; about knife fighting there is no end of controversy. A good book to have a look at to get into the realistic knife-fighting mentality is Don Pentecost's *Put 'Em Down, Take 'Em Out: Knife Fighting Techniques from Folsom Prison* (Paladin Press, Boulder, Colorado, 1988): "Many who thought that they were the "world's deadliest man" laid down when faced with true adversity!" (p. 4) "The traditional martial arts are, in fact, busy perpetuating the most useless shit imaginable." (p. 12) "The martial arts are selling a style. Eastern religion, philosophy, language, tradition, and culture do *not* promote effective self-defense. In fact, most martial artists are neither trained nor proficient streetfighters because, simply put, they do not fight." (p. 13) Modern martial arts have been produced for people largely in urbanized societies, and have thus had a preoccupation with empty-handed techniques because that is all that people have outside of pistol carry, if one is lucky, in the United States. As shown in another classic work, James LaFond, *The Logic of Steel: A Fighter's View of Blade and Shank Encounters*, (Paladin Press, Boulder, Colorado, 2001), most knife encounters are sneak/surprise attacks, where the victim may not even see the blade coming. Street knife attacks in the modern urban jungle are not like the classic standoff portrayed in martial arts books, where the attacker approaches in a ritualistic overhand attack (as in the shower scene in the movie *Psycho* (1960)), but are unexpected, and knives come out of seemingly nowhere, and then your guts are in your hands, then slipping through your bloody fingers, gliding onto the street, down the sewer grate.[257]

That being said, I distinguish between possible knife fighting in the modern urban jungle and the collapsed world to come. As I see it, the surprise element may be much less for true hardcore

---

[257]   http://www.nononsenseselfdefense.com/knifelies.htm.

survivors. In the post-apocalyptic world, being alert and battle ready will be a *sine qua non* of existence. Thus, there are situations where knives could be drawn as last-ditch weapons, possibly if the main weapon has been dropped/knocked out of the hand, or broken. More importantly, the knife in collapse, End of the Rule of Law situations, would be used against unarmed muscle-bound, martial arts-based-attackers (no doubt multiple invaders), something legally out of bounds now in most situations, barring perhaps multiple attackers. Remember, morality has gone down the toilet.

The knife in the hands of a competent fighter is more than enough to take down the trained, unarmed combat person. As Richard Ryan has said: "The hard truth is that you're unlikely to ever control someone who's armed with a knife. You'll never grab or trap the knife hand, you'll never lock or break the arm or wrist, and you'll most certainly never take a knife away from all but the most incompetent attackers."[258] And: "The knife offers no margin for error, so unless your attacker is an idiot, your chances of recognizing, intercepting and controlling him – or his weapon – are about as good as your chances of winning the lottery."[259]

Therefore, what a survivalist would want in such a post-apocalyptic situation as a primary last-ditch melee weapon for up close and personal attacks, is not a tactical folder, or even a superb fighting knife such as the iconic KA-BAR, or otherwise excellent knives produced by Glock with blades about 6 or 7 inches, but *big* knives, not short sword length, but around the 12-14 inches mark.[260] Any longer and one may as well carry a weapon such as a gladius, which in itself is one clear option. This "big knife" philosophy was the position taken by Bill Bagwell, writing in the "Battle Blades" column in *Soldier of Fortune* magazine from 1984 to 1988, and published in his book, *Bowies, Big Knives and the Best of Battle Blades*, (Paladin

---

[258]   R. Ryan, "Against the Odds," in R. Horowitz (et al., eds), *The Ultimate Guide to Knife Combat*, (Black Belt Books, 2007), pp. 239-241, at p. 241.

[259]   As above.

[260]   J. M. Ayres, *The Tactical Knife: Designs, Techniques and Uses*, (Krause Publications, Iola, 2010), *Survival Knives: How to Choose and Use the Right Blade*, (Skyhorse Publishing, New York, 2018).

Press, Boulder, Colorado, 2000). Along similar lines: Paul Kirchner, *Bowie Knife Fights, Fighters, and Fighting Techniques*, (Paladin Press, Boulder, Colorado, 2010), and blade master James Keating: *Gentleman's Steel Reader*.[261] The idea here is simply that if it comes to using a knife, you want to have a badass piece of steel to perforate, chop, slice, dice or otherwise serve up your opponent. Against a smaller blade, while these may technically be faster, the big blade is not slow and has reach and heft. There are numerous suitable Bowies available, made by almost all leading knife firms, such as the ESEE Junglas and Busse Bushwacker Mistress, a range of Bowies from Buck and Cold Steel (e.g. ranging from the premium quality Natchez Bowie to the machete-style Black Bear Bowie). All are of good quality to have in your hand when the chips are down.

The Bowie is not the only style of big blade, with there being the iconic kukri made by almost as many blade firms as those making Bowies. This weapon is spiritually linked to the legendary Gurkha soldiers, and has been used in many battles, an interesting one being when Gurkha Bishu Prasad Shrestha, fought 40 armed men on an Indian train. Armed with his kukri, he killed three, injured eight, and prevented the robbery, according to some reports. He said that he could not have survived without the kukri. While there are numerous manufactures of kukris in the West, especially the United States, one could consider purchasing one straight from the Gurkha homeland of Ex-Gurkha Khukuri House,[262] and/or Everest Blade of Nepal.[263] It is worth noting that in the past Nepalese warriors as a test of manhood would often take on a tiger with their kukri, no mean task, and presumably some survived. The following extract for your reading pleasure is from J. G. Wood, *The Uncivilized Races of All Men in All Countries*, volume II, (1868), just ignore the Victorian snug superiority:

---

[261] http://indigenousability.blogspot.com/2018/01/a-consversation-with-master-bladesman.html; http://www.jamesakeating.com/; http://www.jamesakeating.com/catalg3.html.

[262] https://nepalkhukurihouse.com.

[263] https://everestblade.com.

Perhaps not better proof can be given of the power of the weapon, and the dexterity of the user, than the fact that a Ghoorka will not hesitate to meet a tiger, himself being armed with nothing but his Kookery. He stands in front of the animal (see the next page), and as it springs he leaps to the left, delivering as he does so a blow toward the tiger. As the reader is aware, all animals of the cat tribe attack by means of the paw; and so the tiger, in passing the Ghoorka, mechanically strikes at him. The man is well out of reach of the tiger's paw, but it comes within the sweep of the kookery, and, what with the blow delivered by the man, the paw is always disabled, and often fairly severed from the limb. Furious with pain and rage, the tiger leaps round, and makes another spring at his little enemy. But the Ghoorka is as active as the tiger, and has sprung round as soon as he delivered his blow, so as to be on the side of the disabled paw. Again the tiger attacks, but this time his blow is useless, and the Ghoorka steps in and delivers at the neck or throat of the tiger a stroke which generally proves fatal. The favorite blow is one upon the back of the neck, because it severs the spine, and the tiger rolls on the ground a lifeless mass.

This shows quite well the truth of Philip S. Rawson's statement:

The fundamental appeal of the sword to its owner lies in its serving as an embodiment of his phallic energy, symbolizing his aggressive power as a warrior. Through his sword this power extends even to the dispensation of life and death.

Other big blade options include Asian blades such as the Wing Chun butterfly knives[264] and the Japanese tanto. The iconic tanto need not be discussed in any detail here; it is widely recognized as having superb penetration in stabbing, and massive tissue-destructive potential due to its wicked chisel-shaped edge (in one version), as well as being a fully-adequate slasher. Go for a weapon with a blade length of around 12 inches, whether your choice is the

---

[264]   https://www.everythingwingchun.com/wing-chun-butterfly-swords-s/35.htm.

classic Japanese design, or more modern tactical variations with polypropylene handles.[265]

Kung Fu Butterfly swords, such as the Wing Chun butterfly knives/swords, are used in pairs, fighting with a weapon in each hand. The blade is about as long as one's forearm, which is to provide protection when blocking, the blade is spun using the trapping guards, like a sai, so that the weapon lies along the forearm, edge out to opponent. Thus, some of the commercial products which have flat trapping guards, are difficult to spin, having flat surface that does not permit smooth hand movement. That may not matter if one is not aiming to be a classical expert. Still, a proper set of butterfly swords are relatively expensive, and so for this reason alone, would not be the best weapons choice for ordinary folk. It is a different matter if one has trained in the Wing Chun system, of course.

With all of this, you will be ready for the end. Good luck, because we are going to need it.

CONCLUSION

One of the most dangerous errors is that civilization is automatically bound to increase and spread. The lesson of history is the opposite; civilization is a rarity, attained with difficulty and easily lost. The normal state of humanity is barbarism, just as the normal surface of the planet is salt water. Land looms large in our imagination and civilization in history books, only because sea and savagery are to us less interesting.

–C. S. Lewis.[266]

Western civilization is dying. As Keith Preston observes, "[c]ivilizations die when their elites lose faith in their own civilization to such a degree that the will to survive no longer exists."[267] The

---

[265]   J. M. Yumoto, *The Samurai Sword: A Handbook*, (Tuttle, Tokyo, 1989).

[266]   C. S. Lewis, *Rehabilitations and Other Essays*, (Oxford University Press, Oxford, 1939).

[267]   K. Preston, "The Nietzschean Prophecies: Two Hundred Years of Nihilism and the

situation regarding even this significant existential threat to the West, is as we have seen substantially worse, since the elites, exhibiting pathological altruism at the zombie level,[268] actively are permitting brutal crimes to go unpunished, or unpunished until forced to act, with a collapse of social capital. As Gwendolyn Taunton also laments: "Our civilization is dying ... the chasm of oblivion looms larger than even Spengler predicted for the West. What serves to tie our community together? There are no ties of kindred, no bonds of affection betwixt the masses of faceless individuals that compose our cities – the average man can barely stand to look his neighbor in the eyes. The image of the Nation and the State is shattered beyond repair; we can barely accord our government with credibility little alone trust and respect – thus does our civilization linger on, opening wide the gates of death to permit the vultures to pick our carcass clean."[269] Nevertheless as Troy Southgate has also observed, "chaos and disorder ... inadvertently contains the redeeming elements of sanity and redemption, and this unifying spirit will engender a common identity and enable people to pull together and fight back."[270] That is the upside of going down. However, even if this "come together" scenario of circling the wagons does not occur, we still will have interesting times, as we face Ragnarök and the inevitable death and destruction, of all that oppresses us now. The late Guillaume Faye (1949-2019) was excited:

> Despair is not appropriate. The end of the world is good news, even if it will occur soon with distress and suffering. After the coming shadows will come the light. Human history is far from reaching its end. Preparing for catastrophe and rebirth means transforming one-self *from the inside*. The tragedy on the horizon is perhaps the will of what is called God or fate. We are ruled by

---

Coming Crisis of Western Civilization," in *The Radical Tradition: Philosophy, Metapolitics and the Conservative Revolution*, (Primordial Traditions, 2011), pp. 173-180.

[268] See *Occidental Quarterly* vol. 13, no. 2, Summer 2013, Special Section: White Pathology.

[269] G. Taunton, "Foreword," in *The Radical Tradition* as above, pp. 7-9, at pp. 1, 8.

[270] T Southgate, "Organising for the Collapse," in his *Tradition and Revolution: Collected Writings of Troy Southgate*, (Arktos, London, 2010).

forces which we do not understand and which play dice with us. A new world is about to be born. Man is despairing, but despair is inhuman. The future is thrilling because it is catastrophic. We are dice in God's hands. Who is God?[271]

We are soon going to find out. Hopefully, the material presented above will be an aid in beginning your survivalist education if you are a novice, hence giving you at least a fighting chance.

[271] G. Faye, *Convergence of Catastrophes*, (Arktos, London, 2012).

X. THE ENVIRONMENTAL CRISIS

This chapter was prompted by reading at *American Renaissance,* the article by Philip Santoro, "What does it Mean for Whites if Climate Change is Real?" September 10, 2017.[272]

The point of the article is to examine the hypothetical – if climate change is real, then what is the significance of this for the "White" tribe?[273] Instead of "White," I prefer say, the Northern European/Nordic peoples, but let's get on with it.

There has been much debate within Christianity, with the publication in 1967 of Lynn White's paper, "The Historical Roots of Our Ecological Crisis," *Science,* vol. 155, 1967, pp. 1203-1207. The claim was made that Christianity carried forward a separation between humans and nature, leading to a philosophy of domination, and ultimate exploitation by science and technology. As White said:

> In the history of our culture. It has become fashionable today to say that, for better or worse, we live in the "post-Christian age." Certainly, the forms of our thinking and language have largely ceased to be Christian, but to my eye the substance often remains amazingly akin to that of the past. Our daily habits of action, for example, are dominated by an implicit faith in perpetual progress which was unknown either to Greco-Roman antiquity or to the Orient. It is rooted in, and is indefensible apart from ... Christian theology. The fact that Communists share it merely

---

[272] https://www.amren.com/news/2017/09/climate-change-mass-immigration-green-identity-politics/.

[273] https://www.amren.com/commentary/2017/04/white-environmentalism-roosevelt-conservation-environment/.

helps to show what can be demonstrated on many other grounds: that Marxism, like Islam, is a … Christian heresy. We continue today to live, as we have lived for about 1700 years, very largely in a context of Christian axioms. What did Christianity tell people about their relations with the environment? While many of the world's mythologies provide stories of creation, Greco-Roman mythology was singularly incoherent in this respect. Like Aristotle, the intellectuals of the ancient West denied that the visible world had a beginning. Indeed, the idea of a beginning was impossible in the framework of their cyclical notion of time. In sharp contrast, Christianity inherited from Judaism not only a concept of time as nonrepetitive and linear but also a striking story of creation. By gradual stages a loving and all-powerful God had created light and darkness, the heavenly bodies, the earth and all its plants, animals, birds, and fishes. … And, although man's body is made of clay, he is not simply part of nature: he is made in God's image. Especially in its Western form, Christianity is the most anthropocentric religion the world has seen. … Christianity, in absolute contrast to ancient paganism and Asia's religions (except, perhaps, Zoroastrianism), not only established a dualism of man and nature but also insisted that it is God's will that man exploit nature for his proper ends."

The argument here, though, taken to its logical conclusion, leads to the blame ultimately being placed on science and technology. Again, there are points both for and against scientific and technological advancement, and we do not yet know whether the Western path has more merits than disadvantages. If we bring down civilization and kill ourselves, then, of course, Lynn White critics of modernity will be correct.

In short, it all comes down to the evidence, and my guess is that techno-industrial civilization will destroy itself one way of another.

Back now to the Santoro article on climate change. He is right to point out that today, political agendas are operating through the sciences, including biology and psychology. Climate change, in particular, has been politicized. But, they have done this with everything in environmentalism because in the post-World War II world, people abandoned nature, and have moved to an over-spiritualized and over-civilized view of man.

Apart from possible political bias in climate research, that should have been countered, Christians oppose climate change because it rams home to them that some local, less dramatic version of apocalyptic survivalism could be just around the corner. The Growths oppose climate change mitigation because it is bad for business. Peak oil has similar politics, with many Christian groups siding with Big Oil in denying that there is any impending fuel crisis. Oil comes from the deep earth, abiogenically.[274] Common to all of this is a lack of review of scientific literature and data, to the accepted standards, accepting the fallibility of science, and everything else.

Again, it is all a question of evidence. One needs to read the scientific papers, evaluate them, and if necessary, attempt to replicate experiments, or ascertain new data. In the case of climate change data, this is difficult for lone researchers to do first hand, so often we are left only with the option of scientifically evaluating the published papers. Few of the internet critics of climate change actually spend many hours examining papers to get an informed opinion. It is mainly unthinking responses to climate events in their neighborhood.

All that the global warming hypothesis entails is a rise in the average global temperature since the pre-industrial era. This rise has either occurred or it has not. If it has occurred the cause of the rise is either natural, man-made, or a mixture of both. And, if the cause is real, there may, or may not be, climatic effects. There may or may not be, say, more extreme weather events. All of these things are empirical factual questions which are of enormous scientific complexity. Yet I often see the low IQ arguing that say "mass snow falls in some part of America refute the global warming hypothesis and/or, climate change." Likewise, a severe heat wave alone, does not prove climate change. It could be part of a larger case of evidence

---

[274] https://en.wikipedia.org/wiki/Abiogenic_petroleum_origin.

against the hypotheses, or it could actually confirm them. It all depends upon further bodies of evidence.

All of this evidence is subject to high levels of uncertainty, and open to revision, ideally, but scientists are human rats like us, and hate to see their pet work get refuted, so refutations are much less common than philosophers such as Karl Popper proposed.

Santoro is therefore quite correct in saying:

> If we are not saving the environment for our people, who are we saving it for? The anti-natalism of left-environmentalists—directed mainly towards white countries—amounts to protecting another tribe's future at the expense of ours. We're supposed to save our environment only to turn it over to immigrants? It makes no sense to look after one's patrimony only to give it away to outsiders—which means the entire leftist solution to climate change is bogus.

> Climate change does not have to end in national suicide. The Right's response must seek to combat its effects in the interest of our national well-being. Initiatives to slow or reverse climate change are far less crucial than strengthening our capacity to deal with natural disasters in low-lying areas and to capitalize on new opportunities in the warming north. Assistance to other countries is the lowest priority. Resettling millions of their displaced into our countries is not acceptable. Those leaving the global south would soon be replaced by new births anyway. Migration triggered by climate change would overwhelm us.

Anyone with a survivalist mind set should be prepared to run with the global warming/climate change ideology, and turn the argument to their advantage. Yes, this sort of severe climate change[275] indicates that globalism will lead to ecological disaster, and that Garrett

---

[275] http://nymag.com/daily/intelligencer/2017/07/climate-change-earth-too-hot-for-humans.html.

Hardin's life boat environmental ethics is correct.[276] Hence, every tribe for itself.

However, for most of the Right, especially Christian groups, with their over-spiritualized concept of man, this sort of eco-nationalism/ eco-tribalism, is not explored. They would prefer to be fucked up the anal hole with a pineapple by Big Oil and the corporate fuckers, rather than distinguish themselves from the Left and on the environmental issue.

For all we know, the worst-case environmental doomsday could be just around the corner: collapse by 2030.

Surviving the world of ecological collapse, *Mad Max/Road Warrior* with no feminists or hot cars, will require adopting a completely ruthless Darwinian survivalist Garrett Hardin philosophy of live and let die, which the Christians just don't have the guts to embrace. Nor, for that matter do many of the Right, who still strive for respectability in a time of impending apocalypse. And, of course, everybody deep down loves consumerism, materialism and "civilization," while it all makes Viking Age Barbarian want to puke.

These issues are complex even for PhD educated scientists, and the ordinary guy does not have much chance of working through things: that I grant. So, in the end it will be impossible to reverse climate change, if it is occurring, and the worst will happen anyway, regardless of any carbon cutting done by Americans. Thus, while I personally accept that the worst of climate change is before us, I do not advocate restricting carbon emissions, or shutting down industry. It is far too late. Rather, consume, party and enjoy what life remains, for the human story is coming to an end.

---

[276] http://www.garretthardinsociety.org/articles/art_lifeboat_ethics_case_against_ helping_poor.html.

XI. FROM BLADES TO FISTS

SWORDS & AXES

Articles at the well-respected Arma.org site, seems to be making a physicist's case of the sub-optimal performance of contemporary swords because of intrinsic design flaws.[277]

Even if this is correct, sub-optimal performance is better than hitting someone with your fist, or head-butting them, or cock-whipping them, and modern swords appear to still cut "good enough." I am still impressed by the Cold Steel meat cutting videos, and would dearly like their Viking sword, even though historicists have some criticism of the size of the handle, showing that some people are never bloody happy.[278]

Anyway, considering cost, my preferred melee weapon is the axe; here is a good article on two-handed Petersen type M axes, which have a top cusp/horn for thrusting, and a bottom one for hooking/trapping/down blows.[279]

As James LaFond has noted: "Axes are second only to hammers and malls in dealing with helmets. One defeats a helmet most simply by shocking the neck that supports the head that wears it and knocking that man out. An axe also damages flesh and bone through mail and padding better than a sword, due to immediate leverage

---

[277] http://www.thearma.org/spotlight/GTA/motions_and_impacts.htm#.Wlgvj2iCxaQ; http://www.thearma.org/spotlight/GTA/motions_and_impacts2.htm#.WlgvqGiCxaQ; http://www.thearma.org/spotlight/GTA/motions_and_impacts3.htm#.WlgvwmiCxaQ.

[278] https://www.youtube.com/watch?v=f5rwrUhihEI;https://www.youtube.com/watch?v=-ru8oOpUKqg.

[279] http://sagy.vikingove.cz/two-handed-axes/.

at the point of impact. Axes are fearsome leg-cleaving tools. See the battle of Visby on Gotland for the high frequency of leg cleaves in Norse battle of the mailed era. This is covered extensively in John Keegan's masterful book *The Face of Battle*. In the latter half of the mailed era, Robert the Bruce, famously cleaved a rival knight from the saddle before the battle of Bannockburn. Using the axe in the saddle permitted the rider to rise in the stirrups and then descend with full bodyweight drop, whereas ancient riders without stirrups used pendulum stroking mechanics. Ever after the axe was adored by Scotsmen despite a real lack of timbered lands. The loch-bar axe was used as late as the 1700s in battle against the British, taken down from atop family hearths to repel the invader. Might this have had to do with the habit of the Scots running down upon the invader from the highlands and cleaving downward?" Maybe.[280]

And, here are the noble Nordics Thrand and Eldgrim showing that a little Viking hand axe can easily decapitate and slice open heads.[281]

What more do you want for your shopping dollar?

THE CLUB

The main close-range weapon in the Neolithic age was the wooden club, probably the first human weapon. The spear involved a bit more engineering, and came later. The club could be a big chunk of wood, picked up off the ground, ready to thump anything the grunter did not like.

Archaeologists interested in head injuries exhibited in skulls of over 5,500 years ago, have investigated, using synthetic skulls typically used in wound ballistic tests of firearms, the likely types of injuries.[282]

---

[280]   http://jameslafond.com/article.php?id=11993.

[281]   https://www.youtube.com/watch?v=UGW8pPwmiCA.

[282]   https://www.cambridge.org/core/journals/antiquity/article/understanding-blunt-force-trauma-and-violence-in-neolithic-europe-the-first-experiments-using-a-skins kullbrain-model-and-the-thames-beater/021170E064757BBF7BF3E1870044A60B/.

It was found that Neolithic wooden clubs, such as the "Thames Beater," were highly effective in smashing people's skulls in. This seems to be a lot of work to prove the obvious, since wooden weapons have been used throughout human history, and it is clear that with a piece of solid hard wood, anyone with some muscle, can smash things like bricks and natural stones, which would certainly be as hard, if not harder than a human skull. Anyway, the tests produced shattered skulls similar to skulls held as historical records.

There is a lot on the internet about melee weapons for defense and the like, and whole martial arts systems are based around particular weapons like the katana (Japanese sword). Yet, the simple hunk of wood, available at your nearest tree, provides all the politically correct protection one needs. And, when the threat has gone, you can burn the wood and have a pot of billy boiled tea, which I have cooking away on the camp fire right now.[283]

This last video has a touching end scene, indicating that even shield maidens are gals, deep down.

THE STAFF OF LIFE

A wonderful, simply wonderful article written by James LaFond on the describes merits of the staff as a superior melee weapon against all-comers, provided the opponent is without armor and a shield, and there is adequate space.[284] This was the position of George Silver (1580-1622) in *The Paradoxes of Defence*.[285]

Actually, George was somewhat ambiguous about this, and thought maybe that the forest bill was better in single weapon combat in an open space.

core-reader.

[283] https://www.youtube.com/watch?v=gROkxlT6_dw;https://www.youtube.com/watch?v=34IIoCWqmwo.

[284] https://www.jameslafond.com/article.php?id=9647.

[285] http://www.pbm.com/~lindahl/paradoxes.html;http://jwma.ejmas.com/articles/2001/jwmaart_docherty_0501.htm.

The short staff is most commonly the best weapon of all other, although other weapons may be more offensive, and especially against many weapons together, by reason of his nimbleness and swift motions, and is not much inferior to the forest bill, although the forest bill be more offensive, the short staff will prove the better weapon.

The passage is logically inconsistent, but what the fuck does it matter? Both are fine weapons.

Joseph Swetnam (died 1621) is another badass from that era, and in his treatise *Schoole of the Noble and Worthy Science of Defence*,[286] he is clear about the merits of the staff. Old Joe also wrote a shockingly misogynist treatise, *The Arraigment of Women* (1615), on the allegedly sinful, deceitful, wicked nature of women.[287]

I have trained with the staff for decades now, and in a single melee weapon death match, no armor or shield, adequate moving space, it would be my weapon of choice. I have used the staff in defense against a dog which was set on me, and feral dogs who had killed my calves, and who were foolish enough to turn on me. Paranza Lunga, Sicilian staff fighting, has been practiced for almost a thousand years for defense against humans, and for shepherds, to defend flocks of sheep from wolves.[288]

Staff fighting superiority requires strength by the bucket load. I would advise using the training staff as your primary upper body work out. Start with wood, but work up to thick steel pipe. At my peak, I could lift a slab of the thinner gauge railway iron. I have also constructed staffs with weights on the end, slabs of wood duct-taped together. Of course, a barbell can be used, with a weight at one end.

Unlike swords, where one needs to protect edges, the staff is expendable, and can be replaced by another after the battle.

---

[286]   http://www.thearma.org/Manuals/swetnam.htm#.Whn9tWhL9aQ.

[287]   https://mysite.dmacc.edu/personal/sdphillips2/instructor/lit190/Course%20Materials/Joseph%20Swetnam%20The%20Arraignment%20of%20Lewd.pdf.

[288]   https://www.youtube.com/watch?v=MXJoEuII7jQ; https://www.youtube.com/watch?v=Pr7KJ1tV7Pk.

Smash every fucking thing! Example, my friend Scott, who trains in European Martial Arts (sword primarily) was boasting of his prowess to my, then, 14 year-old nephew. I said, why not do some light sparring with him, you with this short stick to represent your sword, he with a quarterstaff. The quarter staff was tapered to a point at the end, and being olive, the burs at the business end were left. I instructed the kid in the basic six moves of the Wing Chun pole … focus on centerline thrust, cling, stick to the weapon then force/lever through. Use superior leverage to destroy him. They engaged, and the lad thrusted through, moving out the sword/stick, and ramming the staff's point into his hand, ripping off skin from the top of his hand. It was amazing how profusely the wound bled. Most people don't fight on when pissing blood. They are not so tough.

You must train with the staff. It is more important that getting a blow job. The thought police may take your guns, swords, and knives, but they cannot stop you from raping every hardwood tree to secure the fruits of pandemonium.

WITHERING WING CHUN

It was good to see James LaFond giving an expert opinion of Wing Chun's chain/running punch, and relating this back to old school boxing which also used "milling." The discussion was inspired by an informative YouTuber who rightfully shits on Wing Chun, in a respectful bowel movement.[289]

I began training in Wing Chun in the early 1980s, doing boxing and wrestling with my father and his friends before that. The old guys boxed and wrestled in the Great Depression, usually in illegal bare-knuckle matches, that never died out here in the US of A, especially the mighty state of Texas.[290]

---

[289]   https://www.youtube.com/watch?v=47-PcVsRqEU&feature=youtube;   https://www.jameslafond.com/article.php?id=10551.

[290]   https://www.wfaa.com/article/news/nation-world/fight-club-legalized-bare-knuckle-boxing-may-be-next-big-show-in-ring/507-561277946.

When the Bruce Lee craze hit in the 1970s, I did Tae Kwon Do for the kicking, and because, in my deluded teenagerhood, I thought that it was cool, and that I was hot, but was not. Fortunately, the mainland Chinese Wing Chun guy I trained under a bit later, who claimed to have been a student of Bruce Lee (bullshit, I later discovered), had also done other martial arts before settling on Wing Chun to teach, as his Wing Chun was not great. His mate, a Thai, was a reasonable Muay Thai boxer, so all of this was thrown together in the hope that something would come out in the wash. Then, on Friday and Saturday nights, club members did security at a tough bar, so one learnt whether something worked or not.

During a later Wing Chun training time, I had a one-on-one private sparring with a visiting Wing Chun Grandmaster, in private. I was into full-on powerlifting as well as martial arts, coming in at 303 lbs, 600 lbs squat etc. The match was a draw, no-one seemed to be able to get into anyone. However, I probably was holding back out of youthful respect, and may have been able to flatten the little guy if I had lost my temper, or if this was serious shit. Like most kung fu-ers, WC are generally weak, opposed to power training.

About that time some Wing Chun masters did get taken down, literally,[291] an early ground 'n' pound. This incident showed that Wing Chun alone was puny, and one needed skill in ground fighting, as well as plenty of muscle power. Wing Chuners are generally muscularly challenged. Yet, the status quo opposed all of that, still insisting that the fight is over once it goes to the ground, so ground fighting is not in the system. Sure, a ground fight is not one's first preference given all the sharp pointy rubbish on the ground, but too bad if shit does happen and you slip on someone's blood.

Let's not even talk about mystical chi bullshit. Fuck chi with a mystical chilli.

So, my opinion from being in the scene decades ago is that the critics of Wing Chun are largely right. The conventional Chinese kung fu approach to fighting will get you killed on the street. The front-on pigeon-toed training stance is taken by too many ill-trained

---

[291] https://www.youtube.com/watch?v=979JwlCUBdI.

154

practitioners as a fighting stance, which is ridiculous, yet you would be amazed at how many books advocate just that. It shows the lack of training and eager commercialism of many. And, there is not much these guys could do against these monsters.[292]

The running punch is supposed to be a transition to *chi sao*, sticky hands, which then can be a further passage to grappling, but the skill level of most WC is too poor to be able to utilize that. An exception was grappler Larry Hartsell who knew how to integrate grappling into *chi sao*, or was it *chi sao* into grappling?[293]

Regarding weapons, in Wing Chun the butterfly swords and the six and a half poles come at the end of the system, and most masters are pretty terrible at this, since the system is very much for urban environments, and their focus is on teaching the punters unarmed skills, to get the biscuits. Even in the US there are problems with carrying such big knives in some jurisdictions, so the popularity of teaching these weapons, over smaller knives that can be hidden, is not much. Most books treat the pole as an archaic weapon and simply cover it for historical completeness, when in fact, for many rural environments, the pole/quarterstaff is an excellent melee weapon given room to maneuverer. The main books seldom feature how one would use Wing Chun weapons to fight other weapons from other systems, such as say a Viking Dane axe, or Chinese war sword, dadao. The focus is on doing the forms the way master did poo poos.

In conclusion, one would not be adequately trained in melee weapons from most conventional Wing Chun training, although there are sure to be exceptional teachers who are immune to my criticisms. In general, it is far better to go to a Historical European Martial Arts School (HEMA), if you cannot be trained by James LaFond.[294]

---

[292]   https://www.youtube.com/watch?v=rzA8aCl9ZJk.

[293]   https://www.amazon.com/Jeet-Kune-Do-Entering-Grappling/dp/0865680515/ref=sr_1
_1?ie=UTF8&qid=1529897709&sr=8-1&keywords=larry+hartsell&dpID=51ABZP83BVL&p
reST=_SY291_BO1,204,203,200_QL40_&dpSrc=srch.

[294]   https://www.hemaalliance.com.

In the world of MMA, Wing Chun faces extinction, at least in the West, although it is more popular than ever in Hong Kong, becoming something of a symbol of Hong Kong. The system though suffers from the same "classical mess" that Bruce Lee saw in all the Eastern martial arts, including its theory of weaponry. Above all else, many practitioners are mushy, and you can tell this by looking at the rice/carbo guts, long being out of shape. Muay Thai men, and even women, would eat them for a side dish, and a bum wrestler would bum them, sadly. Again, there are exceptions.

Still, this martial art, in my heretical modified form served me well, and is one part of my neo-Viking system, that I personally train in, and never teach. Why teach? My training for survival is of key importance, and there is no time for anything else. I probably should have been training now instead of typing this frog shit. But, everyone wants a sense of closure.

That all said, there may be some things from the classical system worth salvaging, from good old master Wong, who has mastered the art of the profane,[295] as well as Wing Chun red pill lessons about crazy womyn.[296] Master Wong may be Western Wing Chun's last hope, despite his critics.[297]

---

[295]   https://www.youtube.com/watch?v=7sDOc_HFcrc.

[296]   https://www.youtube.com/watch?v=2w-47vchdB4;https://www.youtube.com/watch?v=CYFzy5lz1i4.

[297]   https://www.youtube.com/watch?v=V-lKenZQbKI; https://www.youtube.com/watch?v=hp9-jKO_K98.

PART III

FUCKING MATHEMATICS

XII. MATHEMATICS AND THE COLLAPSE OF CIVILIZATION

Some readers are wishing that I die, so after this article I will oblige them by ceasing to exist, except, perhaps only as an abstract entity. (This refers to the blog article by Dr. A Crank, himself a non-existent entity, whose name was used by myself and the sexy female math geek to publish this shit.)

Let's consider the question of the survival of knowledge after the collapse, the collapse we all talk about and jack off to in our lonely bedrooms, if we are lucky.

There is not much on the mechanism by which people are supposed to rebuild civilization. One school of thought goes for preserving books in hard copy.[298] Useful, no doubt, but there still needs to be people with enough time and IQ to read and understand them.

Lewis Dartnell in *The Knowledge* (2014), thinks that humans can rebuild civilization quickly if the catastrophe which knocks most fuckers off leaves a lot of resources, so that clever grunters can apply fundamental scientific principles in a bootstrap fashion, rebooting civilization by preserving the scientific method.

Ok, this is worth doing, and we do not want to go back to the "bad air" hypothesis of disease after the germ theory and public health measures (e.g. don't shit in or near your water supply), but things are not so easy.

---

[298]   https://www.theorganicprepper.com/books-rebuild-civilization/; http://science.sciencemag.org/content/280/5365/832.

159

Take "my field" of crank mathematics for example. A lot of the more technical advances needed to reboot civilization will depend upon technologies that will require mathematical sophistication, at least calculus, maybe vectors and linear algebra. All of this information is stored in textbooks, except maybe the most recent advances, which are in e-journals, perhaps with paper prints. Certainly, all the mathematics needed for engineering exists in text books. But, if say 90 percent of people die off from a collapse scenario, there will not be many mathematics teachers left. In fact, math geeks are likely to die off pretty quick, being pathetically weak, in my experience. Only philosophers are punier and more pathetic. God, I hate those pissants.

Looking at Dartnell's otherwise informative book *The Knowledge* (2014), I did not find any neat program about how to save mathematical knowledge, and this is surely needed for science to get very far at all. In my copy, the index lists mathematics on page 283. This is a footnote which gives the game away:

> Mathematics is one topic that has not been covered in depth here. Calculations are clearly important for engineering designs, and mathematics is the language for the statement of physical law, but it does not lend itself to the explanations of general principles within the scope of this book.

Gee, that's a pity, since science and technology depend upon math.

Thus, it is most likely that mathematical knowledge, as we now know it will disappear in the coming Dark Age. It may be possible to preserve mathematical knowledge of a level up to the end of high school, such as elementary differential and integral calculus, for one or two generations, but this will be difficult given the stresses of survival and the relatively long time taken for education. Even not having paper to work out problems will limit the recovery of knowledge. Topics like projective geometry, surreal numbers, advanced transfinite set theory, category theory, and the mathematics of theoretical physics such as string theory, will be lost, almost overnight. What the fuck will it matter anyway? I wasted my life playing with abstract entities, when I should have been playing with juicy pussy. Now, I am but a decaying circus clown, more nothingness than being.

You will not be able to focus on that math problem when the warlords of the apocalypse come a-killing at your door? You can count on it.

XIII. ON MATHEMATICAL BULLSHIT

There is no question that mathematics is, after a certain level, just a game with symbols that an elite priesthood jack off to. The financial elite have their money; the mathematical elite, symbolic structures. Hence, while it is true that the series 1+2+3+... diverges in the standard sense, mathematicians have found it amusing, and useful for shonky physicists working in areas dripping in infinities, such as string theory and quantum field theory, to come up with a useful result.

Thus, the series 1-1+1-1+..., diverges in the standard sense, but a Cesaro summation can be given, defined as the limit of the arithmetic mean of the partial sums of the series. Then 1-1+1-1+... =1/2, the value which the series oscillates around. See: G. H. Hardy, *Divergent Series*, (Clarendon Press, Oxford, 1949).

But, not all series can be given a Cesaro sum, such as 1+2+3+..., and that is where the Euler/Riemann zeta function comes in. Zeta function regulation involves defining the Riemann zeta function for all values of z as the analytic continuum, which means extending the definition of the function from reals to functions of complex numbers. Thus:

$\xi(x) = S(x) = 1 + 1/2x + 1/3x + ...$

If x= -1 is put into this function, then one gets $\xi(-1) = -1/12$.

Some may conclude that 1+2+3+..., therefore equals -1/12. This, however is incorrect, because strictly speaking divergent series in the standard sense can be assigned a number of different values based on analytic continuum, so the "=" is not used in a standard sense at all. For example, the series 1-1+1-1+..., could be assigned the value 1/3. One can obtain some truly bizarre results such as:

1+1+1+...= -1/2

1+4+9+... =0

Terry Tao discusses these in his paper, "The Euler-Maclaurin Formula, Bernoulli Numbers, the Zeta Function, and Real-Variable Analytic Continuation," April 10, 2010.[299] Tao sees a problem here:

> Clearly, these formulae do not make sense if one stays within the traditional way to evaluate infinite series, and so it seems that one is forced to use the somewhat unintuitive analytic continuation interpretation of such sums to make these formulae rigorous. But as it stands, the formulae look "wrong" for several reasons. Most obviously, the summands on the left are all positive, but the right-hand sides can be zero or negative.

A "problem," indeed!

Even given the zeta function result, if "=" is given its standard meaning then we have a contradictory result, precisely because the summands on the left are positive and non-zero, but the right-hand side is zero or negative, which gives a classical contradiction p&~p. But, if "=" is given a non-standard meaning, then the physicists can't use this reasoning in their shonky physics, to avoid infinites.

It is known, but often ignored, that with bracket manipulation within an infinite series, and with the rearrangement of terms, inconsistencies can be generated, even for convergent series, so that one can get a sum, which is any real number, with a bit of ingenuity. Thus, although 1-1+1-1+...= ½, via the analytic continuation of the zeta function, if one adds brackets, one can get;

---

[299]   https://terrytao.wordpress.com/2010/04/10/the-euler-maclaurin-formula-bernoulli-numbers-the-zeta-function-and-real-variable-analytic-continuation/.

(1-1)+(1-1)+(1-1)+...=0+0+0+0=..., which is not 1/2. My guess is it is 0, or should be 0.

With bracket rearrangement, one can "prove" that 1-1+1-1...=1/2, even without use of the zeta function. Let S=1-1+1-1+..., then:

S= 1-(1-1+1-1+1-...) = 1-1+1-1+... =S. So, 2S=1, or S=1/2.

Consider again, $S = 1 + 2 + 3 + 4 + 5 + \cdots$. We have $-3S = (1 - 4)S = (1 + 2 + 3 + 4 + 5 + \cdots) - 2(2 + 4 + 6 + \cdots) = 1 - 2 + 3 - 4 + 5 - \cdots = 1 - (2 - 3 + 4 - 5 + \cdots) = 1 - (1 - 2 + 3 - 4 + 5 - \cdots) - (1 - 1 + 1 - 1 + \cdots) = 1 + 3S - 1/2$, and we conclude that $-6s = 1/2$, that is $1 + 2 + 3 + 4 + 5 + \cdots = -1/12$.

Here is another so-called "proof":

Term-by-term summation used in Numberphile's video

	S	=	1	+ 2	+ 3	+ 4	+ 5	+ 6	+ 7	+ 8	+ ...	= ?
	S_1	=	1	- 1	+ 1	- 1	+ 1	- 1	+ 1	- 1	+ ...	= 1/2
	S_2	=	1	- 2	+ 3	- 4	+ 5	- 6	+ 7	- 8	+ ...	
	$2S_2$	=	1	- 2	+ 3	- 4	+ 5	- 6	+ 7	- 8	+ ...	
				+ 1	- 2	+ 3	- 4	+ 5	- 6	+ 7	+ ...	
		=	1	- 1	+ 1	- 1	+ 1	- 1	+ 1	- 1	+ ...	= 1/2
S -	S_2	=	1	+ 2	+ 3	+ 4	+ 5	+ 6	+ 7	+ 8	+ ...	
	-		1	+ 2	- 3	+ 4	- 5	+ 6	- 7	+ 8	+ ...	
		=	0	+ 4	+ 0	+ 8	+ 0	+ 12	+ 0	+ 16	+ ...	= 4S

S - 1/4 = 4S ⇒ S = - 1/12

But as a counter to Numberphile, consider this:

1+2+3+4+... = 1+(1+1)+(1+2)+(1+3)+... = (1+1+1+1+...) + (1+2+3+4+...).

Therefore -1/2 + -1/12 = -1/12, which implies that -1/2 =0, which implies that 1=0, an absolute inconsistency.

Clearly, then, all of the above results are rubbish and thus one should be a caution about regarding infinite series as things which can be manipulated according to the mathematical rules normally applied to finite sums. However, noting this leads to some even more

165

interesting results when we re-examine the issue about .999… being equal to 1, in my next arousing paper.

XIV. WHY 0.999… IS NOT EQUAL TO 1!

D ear friends, this may get a bit complex, so eat a big steak and piles of eggs, have a few beers, and I am sure it will go down smoothly. If anyone in class needs to jack off first to relax, please do it now. Use hand sanitizer to kill off the germs, which none of us want these social distancing days, so keep your fucking distance. Ok, let's get started.

Abraham Fraenkel, in *Abstract Set Theory*, (North-Holland, Amsterdam, 1976), gives the following auxiliary theorem from the theory of decimals:

> Every positive (non-zero) real number A, has one and only one expansion into an infinite decimal $A = m.c_1c_2c_3\ldots c_k\ldots$, where m is a non-negative integer and the digits c_k assume the values 0, 1, 2, 3, …9, with the proviso that after every c_k digits different from 0 will appear. If from a certain place and onwards in the decimal expansion only zeros appear, the decimal expansion is said to be terminating; if otherwise, non-terminating. Therefore, two infinite decimals which are not identical represent different real numbers. (p. 51)

Fraenkel then proposes that a positive real number that can be expanded into a terminating decimal:

(F1) $m.c_1c_2c_3\ldots c_n$, c_n not $=0$

is equal to the infinite decimal:

(F2) $m.c_1c_2c_3...(c_n-1)99999....$

or if it is a positive integer m, to (m-1).9999...

Fraenkel recognized that this claim was surprising, but said that the equality 1=0.999..., can be obtained by multiplying by 9 the equality $1/9 = 0.111...$. (p. 4) See also, G. H. Hardy and G. M. Wright, *An Introduction to the Theory of Numbers*, (Clarendon Press, 1960).

The notion that non-terminating decimals with a repetition of digits, such as 0.999...=1, and 1/3= 0.333..., and so on, are generally accepted, outside of non-standard analysis, to be correct, not only because of the definition given above, but because of a number of mathematical arguments, which will be reviewed and refuted here. For the orthodox positions see "Why is 0.999...Equal to 1?";[300] M. A. Navarro and P. P. Carreras, "A Socratic Methodological Proposal for the Study of the Equality 0.999...=1," *The Teaching of Mathematics*, vol. 13, 2010, pp. 17-34; R. Pemantle and C. Schneider, "When is 0.999...Equal to 1?" (April, 2007).[301]

From the perspective of non-standard analysis, it has been maintained that 0.999...< 1. In general, the number .999..., is thought of as an infinite decimal, and although some reject that idea (including me), many do not: "What is Wrong with Thinking of Real Numbers as Infinite Decimals?"[302] The standard real number 0.999... would be represented by:

.999...; ...999... ,

Which falls short of 1 by an allegedly infinite amount, 1/10.[H] See K. U. Katz and M. G. Katz, "When is .999... Less than 1?" *Montana Mathematics Enthusiast*, vol. 7, no. 1, 2010, pp. 3-30, at p. 5.

In Lightstone's notation:

[300] http://www.quora.com/Why-is-0999-1dots-equal-to-1.

[301] https://www.math.upenn.edu/~pemantle/papers/nearly.pdf.

[302] https://www.dpmms.com.ac.uk/~wtglo/decimals.html.

.999… ; …999… < 1. See: Katz and Katz, above, p. 6; A Lightstone, "Infinitesimals," *American Mathematical Monthly*, vol. 79, 1972, pp. 242-251. For further literature on this point see K. U. Katz and M. G. Katz, "A Strict Non-Standard Inequality .999…< 1," at arXiv: 0811.0164v8 [math.HO] February 24, 2009; K. U. Katz and M. G. Katz, "Zooming in on Infinitesimal 1-.9… in a Post-Triumvirate Era," *Educational Studies in Mathematics*, vol. 74, 2010, pp. 259-273; D. Tall, "Intuitions of Infinity," *Mathematics in School*, May, 1981, pp. 30-33.

Numbers such as H = 0.000… ; 999…, are not members of the hyper-reals, because if on assumption H was, it would be less than any positive real, as it has only os to the left of the ";", which would make it an infinitesimal. But, if an infinitesimal is added to H, there would be a carry over to the left of the ";" so that H would have a non-zero standard part. Hence, the theorem that the sum of two infinitesimals is an infinitesimal would be violated. See; "Is .999…=1? A Non-Standard view."[303] But, I believe that rather than saying that H is not a hyper-real, I prefer to see this as a counter-example, and see the rejection of H as a hyper-real as ad hoc. This is because I reject non-standard analysis as I have argued in my "Fuck Infinity" series.

J. Benardete in *Infinity*, (Clarendon Press, 1964), p. 279, said: "The intelligibility of the continuum has been found—many times over—to require that the domain of real numbers be enlarged to include infinitesimals. This enlarged domain may be styled the domain of continuum numbers. It will now be evident that .999… does not equal 1 but falls infinitesimally short of it. I think that .999… should indeed be admitted as a *number* … though not as a *real* number." He also argued against the proposition that the rational number 2/3= .666…, in his paper, "Continuity and the Theory of Measurement," *Journal of Philosophy*, vol. LXV, 1968, pp. 411-430, at p. 429:

I think that it should be fairly evident that in the theory of absolute continuity the decimal number .6666 … is smaller than

---

[303]   http://www.cut-the-knot.org/WhatIs/Infinity/999.shtml.

the rational number 2/3. We may establish that result as follows. Imagine that there is a road that is exactly 2/3 miles long from its initial to its terminal point. Imagine further that the road terminates in a wall. Then, if one is to walk the full distance of 2/3 miles starting from the initial point of the road, one must come smack up against the wall. Now posit a sheet of paper so thin that, for any arbitrary rational number r, the sheet can be shown by measurement to be less than r inches thick. If one is tempted to regard this sheet as being 0 inches thick, i.e., as having no thickness at all, we have only to stack seven such sheets of paper one on top of the other. The whole stack must be thicker than any one sheet taken separately. Each sheet of paper, and the whole stock taken collectively, must then be ε inches thick, where ε is some actual infinitesimal. Let us now paper the wall with these seven sheets, again laid one upon the other. In walking the full distance of 2/3 miles from the initial point of the road it will not be sufficient merely to come into contact with the outermost sheet of paper. One must penetrate through all seven sheets so as actually to come smack up against the surface of the wall. But suppose one walks only .6666 ... miles from the initial point. That is to say, one walks 6/10 miles + 6/100 miles + 6/1,000 miles etc. ad infinitum – no more and no less. In this case one will not even penetrate the outermost sheet of paper, much less any of the others. We must then conclude that 2/3 is greater than .6666.... .

This could be taken as one counter-argument to the approach which argues that if 1/3=0.333..., then 1=0.999..., because this result is obtained by multiplying the first equation by 3. The mathematical skeptic would need to deny that 1/3= 0.333... .

Another elementary argument for .999...=1, is to let K=.999..., multiply K by 10, giving 9.999..., so 9K=9, and K=1. The problem, as will be seen with all standard proofs, is that this argument begs the question: R. Rucker, *Infinity and the Mind*, (Harvester Press, 1982), p. 79: "this argument overlooks the fact that the difference between 10K and 10 is ten times as great as the difference between K and 1" as there is a "residual infinitesimal quantity below that does not get canceled out."[304]

The mathematical skeptic could also argue, as Fred Richman, "Is 0.999…=1?" *Mathematics Magazine*, vol. 72, 1999, pp. 396-400, points out as devil's advocate, that 9K is 8.999…, not 9, and 8.999…, is not equal to 9. The skeptic could say, that K cannot be canceled, so subtraction of real numbers is not always possible. (p. 396)

John Gabriel, "The 0.999…=1 Fallacy," rightly rejects, on finitist grounds, multiplying the "number" .999…, as it has an ill-defined infinite sequence of digits, 999… . Moreover, the multiplication argument is circular, as one can just as easily argue: let K=0.999…, so 10K=10 x 0.999… . Then 9K=9 x 0.999…, so K=0.999… . The "proof" is equally useless, assuming what has to be proved. If you don't like John Gabriel, then you can find others who recognize this skeptical argument. Dr. Peter Cotton, who plugs for the Cauchy sequence argument (refuted below), and is thus not a skeptic, says that the mathematical skeptic rejects proofs that 0.999…=1, because they are circular, "they assume a calculus on infinite decimal expansions (without first constructing it… which will involve defining 0.999… and 1 to be equal and hence be circular.)"[305] He says that the difference between 0.999… and 1 is arbitrarily small, and that the only standard arbitrarily small number is zero. I disagree, and refute this below.

Yet another possibility is that even if this proof is accepted, there could still be independent arguments for 0.999…< 1, even given arguments for 0.999…=1, in a kind of neo-Kantian antinomy of reason. So, real number theory is negation inconsistent. Such situations of proof "over-determination" are generally ignored by mathematicians, who simply assume that the whole game is consistent, and sound, thus begging the question of the soundness of mathematics. The logical paradoxes indicate that this is not so, and that there can be sound arguments both for and against a proposition. Thus, even a proof that 0.999…=1, is not necessarily a refutation of 0.999…<1, because consistency is presupposed.

[304] http://www.quora.com/why-is-0-999-Idots-equal-to-1.

[305] https://www.quora.com?Why-is-0-999-Idots-equal-to-1.

Another popular argument is based on the limit argument, which is found right across the internet. Here it is argued that "0.999..." is "shorthand" for the limit of the sequence of real numbers, 9/10 +9/100 + 9/1,000 +...., which converges to 1. Therefore, 0.999... and 1 are equal. See for example many pages on quora.com; P. Eisenmann, "Why is it Not True that 0.999...<1?" *The Teaching of Mathematics*, vol. 11, 2008, pp. 35-40. We have already seen in the previous article, that there is a need for great caution in identifying the limit of a sequence with the sequence itself. Richman (as above, p. 397), says: "some distinction between convergence and equality in the present case might be appropriate." D. Tall, cited from Katz and Katz (*Educ. Stud. Math*, paper, as above, pp. 260-261), said: "the infinite decimal 0.999...is intended to signal the limit of the sequence 0.9, 0.99, 0.999... which is 1, but in practice it is often imagined as a limiting process which never quite reaches 1."

Carl B. Boyer, *The History of the Calculus and its Conceptual Development*, (Dover Publications, New York, 1949), said; "Cauchy had stated in his "Cours d' analyse," that irrational numbers are to be regarded as the limits of sequences of rational numbers. Since the limit is defined as a number to which the terms of the sequence approach in such a way that ultimately the difference between this number and the terms of the sequence can be made less than any given number, the existence of the irrational number depends, in the definition of limit, upon the known existence, and hence the prior definition, of the very quantity whose definition is being attempted. That is, one cannot define the number $\sqrt{2}$ as the limit of the sequence 1, 1.4, 1.41, 1.414, ..., because to prove this sequence has a limit one must assume, in view of the definitions of limit and convergence, the existence of this number as previously demonstrated or defined. Cauchy appears not to have noticed the circularity of the reasoning in this connection, but tacitly assumed that every sequence converging within itself has a limit." (pp. 281-282) The same circularity objection can be made against the argument using an infinite series to allegedly show that 0.999...=1.

Real numbers may be defined as the limit of Cauchy sequences of rational numbers. Thus, if x_n and y_n are two Cauchy sequences, then they are equal if the sequence $x_n - y_n$ has limit 0. Here (1-0, 1-9/10, 1-99/100, ...) = (1, 1/10, 1/100, ...) would have limit 0.

Another argument is the Dedekind cut argument. A real number R is defined as the infinite set of all rational numbers less than R. Hence, the real number 1 is the set of all rational numbers less than 1. The number 0.999... is the set of rational numbers r such that: r<0 or r<0.9, or r<0.99... and r is less than a number 1 − (1/10).[n] Therefore, arguing along these lines, every element of 0.999... is less than 1. Hence, it is an element of the real number 1. An element of 1 is a rational number a/b<1, so a/b<1 − (1/10).[b] As 0.999... and 1 contain the same rational number, they are allegedly the same set, so 0.999...=1.

Richman rightly observes that the mathematical skeptic will reject both of these arguments. A definition is set up to rule out the existence of distinct numbers 0.999... and 1, as 0.999... is the cut $\{x \, \varepsilon \, D: x < 1\}$ and 1 is the cut $\{x \, \varepsilon \, D: x \le 1\}$. Dedekind said that these cuts were "only unessentially different" (p. 398, Richman, as above), which begs the question, because, the signs "<" and "≤" have different meanings, and contrary to Dedekind, the cuts are essentially different. Richman says, "in the traditional definition of the real numbers, the equation 0.9...=1 is built in from the beginning," and anyone "who challenges that equation is, in fact, challenging the traditional formal view of the real numbers." (p. 399) As the traditional view of numbers is what is being challenged by the mathematical skeptic, it is obviously question begging to use that account, undefended, in defense.

Consider numbers to be formal strings of symbols. Thus, 1 is denoted by "1." A number such as .999..., consists of the concatenation of signs "9" followed by "9," and so on. Further, there is no good reason why one cannot start at the right hand-side of the number, and write .999..., as .999...999. Mathematics is, after all, just a formal game and one can do what one likes, bar absolute inconsistency. In a sense, one could "terminate" a conventional non-terminating decimal because we know, by definition, that .999... is only going to have 9s in its sequence. This will not be so for say irrational numbers, such as √2, or transcendental numbers such as e and π. But, for all practical purposes, it would not matter what number, if any, was put as a "last' number, or even a special disjunctive last place could be defined; "0 v1 v 2 v 3 v 4 v 5 v 6 v 7 v 8 v 9." Or, in the alternative, there may be no last digit characterizable at all. It matters not in a game with symbols.

Numbers, then, are constructed using the concatenation of signs "0," "1," "2," "3," "4," "5," "6," "7," "8," and "9." The first positive non-zero real number can be defined as:

0.000...001, and the next: 0.000...002, and so on. The number 0.999... is thus, 0.999...999. This number is 0.000...001 less than 1.000...000. One can add 0.000...001 to 0.999...999, to get 1.000...000. Whether one calls 0.000...001 an infinitesimal or not, is moot. It is obtained from an elementary rethinking of the way we view real numbers. Given all of the bs that we have looked at in this series of articles, this approach makes as much sense as anything else.

Interestingly enough we can define the largest real number. It is:

...999.9999...999, an infinite string of 9's going left and right. The reason this is the largest number, is that if one attempts to add to it, one gets a smaller number. There is no way of symbolically representing a larger number.

As well, there will be an infinity of numbers, with no number between them. Thus, it is not true for all real numbers that if x, y, if x < y, then (x + y)/2 < y, as that average number may not be defined, if they are next to each other, so to speak.

It has also been objected by Ted Alper,[306] that there will be difficulties for this theory, for the usual arithmetic operations. To take addition, for example, he asks: what is: x = 0.4999... + 0.4999...? His suggestion is 0.999...998. He says that for arithmetic whenever a < b, then a + c < b + c. So we have: 0.4999... 999 < 0.5. Then:

---

[306]   https://quora.com/why-is-0-999-Idots-equal-to-1.

$$x = 0.4999\ldots + 0.4999\ldots(= 0.999\ldots998) < 0.5 + 0.4999\ldots =$$
$$0.999\ldots999.$$

But, he says, x will need to be some number less than 0.999...999, but greater than 0.4999... + 0.4999... = 0.999...998. There is no such number capable of representation, because nothing falls between 0.999...998 and 0.999...999. However, the above inequality is correct, for x = 0.999...998, which is less than 0.5 + 0.499...999 = 0.999...999. Hence, the objection collapses. But, even if it did not, one could maintain that the usual rules of arithmetic do not mechanically apply to infinite numbers, which is what mathematicians say in other areas. Thus, the rule above would fail for some numbers, such as the last and second to last numbers.

Philosophically this approach should be a joy to mathematical skeptics, iconoclasts and cranks, as it gives a way of refuting Cantor's diagonal argument (the diagonal number, by definition will be in the infinite sequence of concatenated numbers). Further, it gives an example of a number speculated by Graham Priest, to be the largest possible number. See G. Priest, "Inconsistent Models of Arithmetic II: The General Case," *Journal of Symbolic Logic*, vol. 65, 200, pp. 1519-1529. However, where I disagree with Priest is that he seems to think that this last number is inconsistent, while I see no reason for doing that. In fact, adding to the largest number $\Omega = 9999999\ldots$ produces a smaller one, so $\sim(\Omega = \Omega + 1)$.

It is concluded that 0.999... is not equal to 1. All of the standard proofs that 0.999...=1, are circular. Mathematics is in flames, but looking at the dreamy universities, with the masters of the symbolic arts walking around with their hands on their tools (gender neutral tools), you would not think so. The cranks in their mom's basement, doing math and jacking off to porno are right.

XV. THE PARADOXES SERIOUSLY FUCK MATH

What an arrogant bunch of pricks, little ones, mathematicians are, me excluded (not arrogant; big prick). Let's deflate these cock suckers some more with utterly stunning cranky logic.

The existence of logico-semantical paradoxes is a real problem for their paradigm. Mathematicians pride themselves on logical rigour beyond all things, but for over one hundred years, mathematics has faced foundational paradoxes. Before that, going back to ancient Greece, there was awareness of semantical paradoxes such as the Liar.[307]

Here is a standard statement of the paradox from the above source:

> Consider a sentence named 'FLiar', which says of itself (i.e., says of FLiar) that it is false.
>
> FLiar: FLiar is false.

This seems to lead to contradiction as follows. If the sentence 'FLiar is false' is true, then given what it says, FLiar is false. But FLiar just is the sentence 'FLiar is false', so we can conclude that if FLiar is true, then FLiar is false. Conversely, if FLiar is false, then the sentence 'FLiar is false' is true. Again, FLiar just is the sentence 'FLiar is false', so we can conclude that if FLiar is false, then FLiar is true. We have thus shown that FLiar is false if and only if FLiar is true. But, now,

---

[307]   https://plato.stanford.edu/entries/liar-paradox/.

if every sentence is true or false, FLiar itself is either true or false, in which case—given our reasoning above—it is both true and false. This is a contradiction. Contradictions, according to many logical theories (e.g., classical logic, intuitionistic logic, and much more) imply absurdity—triviality, that is, that every sentence is true."

Fine, you may say, simply deny that all sentences are either true or false, the principle of bivalence. Many have done that, but there are strengthened Liar paradoxes that escape that method of defense, using the same source reference:

Consider a sentence named 'ULiar' (for 'un-true'), which says of itself that it is not true.

ULiar: ULiar is not true.

The argument towards contradiction is similar to the FLiar case. In short: if ULiar is true, then it is not true; and if it is not true, then it is true. But, now, if every sentence is true or not true, ULiar itself is true or not true, in which case it is both true and not true. This is a contradiction. According to many logical theories, a contradiction implies absurdity—triviality."

The Liar paradox can also be presented in more complex forms that escape attempts to solve the strengthened versions, such as Yablo's paradox, where there is use made of a list of sentences where reference is made to the sentence being not-true further down the list.[308]

S1: S2: S3: Sn: For all $m>1$, Sm is false.

For all $m>2$, Sm is false.

For all $m>3$, Sm is false.

$$\vdots \quad \vdots \quad \vdots \quad \vdots \quad \vdots$$

For all $m>n$, Sm is false.

$$\vdots \quad \vdots \quad \vdots \quad \vdots \quad \vdots$$

This also generates a paradox.

In the Middle Ages logicians, who were also theologians, debated the meaning of various "insolubilia," such as the Liar, but also other, perhaps more challenging paradoxes, such as the paradox of validity, attributed to "Pseudo-Scotus," someone who was not John Duns Scotus, but whose work ended up in a publication by John Duns Scotus. Pseudo probably lived around the 1340s or 1350s. Here is a version of his paradox:

> Pseudo-Scotus does not realise he is dealing with a paradox. He presents it as an argument against a certain definition of valid consequence: that it is impossible for things to be as the premise signifies without being as the conclusion signifies. Consider the argument: God exists. So, this argument is invalid. Call the argument, π. If π were valid, it would be a valid argument with a true premise, so the conclusion would be true, that is, π would be invalid. So, by reductio, π is invalid. Now the conclusion of π is necessary, since we've just inferred it from the necessary truth that God exists. By the above definition, any argument with a necessarily true conclusion is valid. So, by the definition an invalid argument is valid.
>
> But π is, in fact, paradoxical. For by Pseudo-Scotus' own admission, we've deduced the conclusion of π from its premise. First, we assumed π was valid. Then taking it that the premise of π was true, we inferred that its conclusion was true, that is, π was invalid. So, by reductio, π is invalid, assuming that its premise is true. So, on any account, π is valid, since its conclusion follows from its premise. Hence π is both valid and invalid ... since π is valid, its conclusion is false, so its premise must be false too, that is, there is no God. Again, we could use this argument to disprove anything.[309]

The conclusion that there was no God was taken in the day as showing that the argument was absurd. A more modern version of this argument is as follows.

Consider:

309   https://hesperusisbosphorus.files.wordpress.com/2012/03/istanbul-paradox-hdt-3.pdf.

(A) 1=1

Therefore,

(B) This argument (i.e. (A) → (B)) is invalid.

Now, to use an argument by Stephen Read (*Synthese*, vol. 42, 1979, pp. 265-274, at p. 267), if this argument is valid, it has a true premise and a false conclusion, as every argument with a true premise(s) and false conclusion is invalid, failing to transmit truth. Therefore, this argument is invalid. Hence, the argument (A) → (B) is valid if it is invalid. So, by *reductio ad absurdum*, the argument is invalid. However, taking an alternative track, 1=1 is mathematically true, and hence a necessary truth. It is a principle of most modal logics that what deduced from a necessary true proposition is necessarily true. Then, as premise (B) is deduced from the necessary truth 1=1, then the argument (A) → (B) is valid. So, the conclusion (B) is a necessary truth, namely it is a necessary truth that the argument is invalid. Hence the argument is valid and not-valid, a contradiction!

There are even worse paradoxes that were uncovered in the 20th century, where by the fundamental principles of logic, one can prove any arbitrary proposition, p:[310]

"Lob's argument shows that the use of negation is not needed for the proof that every statement is true. For, let B be any sentence of the language. Create a sentence A such that A is true if and only if it implies B, i.e., (2) A if and only if (A→B). Then argue as follows. Suppose (3) A, then (4) A →B and (5) B. In other words, withdrawing the assumption (3), (6) A →B, i.e., (7) A, so (8) B!"[311]

There are many responses to this triviality proof, but also many rejoinders restating the paradox. It seems that some fundamental logical principles must be given up, but what? Everything seems to be basic and undeniable.

---

[310]   https://plato.stanford.edu/entries/curry-paradox/.

[311]   https://link.springer.com/content/pdf/10.1007%2FBF00245920.pdf.

I do not have a formal answer to give here, as I believe that one cannot ever be forthcoming. Every logician has a different solution, and these solutions all have specific defects. This to my warped mind shows that human reason is basically shit, limited in its capacity to understand the universe, and even to think, because we evolved to primarily drink our enemies' blood from skulls, then die and rot.

The so-called educated like to think that they have some divine right to piss and shit on people like you. But, when it comes down to it, most of higher education is just bullshit. Here's a book full of all the stats about this, including the fact that most graduates basically remember nothing much of their courses after the exam.[312] I am one of the least-educated people on the planet, and I am here to tell you that most of what we learn, in every field is just bullshit, lies, fraud, deception, or simply false, but don't take my word for it; here is a much more respectful position:

There is increasing concern that most current published research findings are false. The probability that a research claim is true may depend on study power and bias, the number of other studies on the same question, and, importantly, the ratio of true to no relationships among the relationships probed in each scientific field. In this framework, a research finding is less likely to be true when the studies conducted in a field are smaller; when effect sizes are smaller; when there is a greater number and lesser preselection of tested relationships; where there is greater flexibility in designs, definitions, outcomes, and analytical modes; when there is greater financial and other interest and prejudice; and when more teams are involved in a scientific field in chase of statistical significance. Simulations show that for most study designs and settings, it is more likely for a research claim to be false than true. Moreover, for many current scientific fields, claimed research findings may often be simply accurate measures of the prevailing bias.[313]

---

[312] https://www.amazon.com/Case-against-Education-System-Waste/dp/0691174652.

[313] http://journals.plos.org/plosmedicine/article?id=10.1371/journal.pmed.0020124.

Most published scientific material is false. Well, so much for the Enlightenment myth of knowledge and truth; if there were no technological things to play with most people would have no use for science at all, and would return to black magic and sorcery.

XVI. THE PEBBLES OF FINITUDE:

The Inconsistency of the Principle of Mathematical Induction

My conceptual rampage today will involve attacking a fundamental elementary mathematical principle, so grab a beer, or eight, and sit down for another arousing read. The blood of mathematics will be splattered everywhere. Barbarians, all enjoy!

The principle of mathematical induction, which is actually a deductive principle, is used extensively in number theory, and it would seriously fuck up mathematics if there were sound objections to it. Here, I argue that there are, and that this knee-caps mathematics.

The principle of mathematical induction is a proof method for proving that a proposition P(n) holds for all natural numbers 0, 1, 2, 3, ... It is like a series of dominoes knocking the next over. And, it is said to work because the axiom of induction is one of the Peano axioms for arithmetic:

If φ is a unary predicate such that:

$\varphi(0)$ is true, and

for every natural number n, $\varphi(n)$ being true implies that $\varphi(S(n))$ is true,

then, $\varphi(n)$ is true for every natural number n.

183

To conduct a proof, first prove the base step, P(0), that the property holds for the number 0, or 1. The second step is the inductive step where one makes the assumption that P(k), is true for some k. Then one considers P(k+1), and shows that P(k) implies P(k+1), thus concluding that the proposition is true by the principle of mathematical induction.

Fine, but is the principle correct? We know from "Wang's paradox," that the principle fails for vague predicates like "small."[314] 0 is a small number; if P(k), is assumed to be small, then P(k+1) is also small (as there is no radical jump in numbers from small to big), so all numbers are small! Philosophers and logicians find this interesting, but mathematicians usually tell you to go away if you want to talk about this, and may even flex their 1-inch biceps to show that they mean business. It is a terrifying sight. Fortunately, they are not likely to drop their daks (apparently derived from "Dad's Slacks"), as there would not be enough microscopes to go around.

Anyway, there are some technical issues raising doubt about the principle of mathematical induction.

Princeton University mathematician Edward Nelson (1932-2014), who once thought that he had a proof of the inconsistency of arithmetic,[315] but was wrong on a technicality (can it be patched up as Andrew Wiles' proof was?), in his book, *Predicative Arithmetic*, (Princeton University Press, Princeton, 1986), expressed doubt about the principle of mathematical induction, saying:

That the reason for mistrusting the induction principle is that it involves an impredicative concept of number. It is not correct to argue that induction only involves the numbers from 0 to n; the property of n being established may be a formula with bound variables that are thought of as ranging over all numbers. That is, the induction principle assumes that the natural number system is given. A number is conceived to be an object satisfying every

---

[314] http://www.jamesrmeyer.com/paradoxes/wang-paradox.html; http://www.univie.ac.at/constructivism/journal/articles/7/2/141.vanbendegem.pdf.

[315] https://golem.ph.utexas.edu/category/2011/09/the_inconsistency_of_arithmeti.html.

inductive formula; for a particular inductive formula, therefore, the bound variables are conceived to range over objects satisfying every inductive formula, including the one in question.

Thus, the principle of mathematical induction is up to its neck in metaphysical assumptions about infinity.

Another problem with so-called proofs by the principle of mathematical induction is that all such proofs assume that arithmetic is not ω-inconsistent, where for some formula A(x), each formula of the infinite sequence A(0),...,A(n),..., and the formula $\neg\forall x A(x)$ are provable, where 0 is a constant of the formal system signifying the number 0, while the constants n are defined recursively in terms of $(x)'$, denoting the number following directly after x: $n+1=(n)'$. Thus, if arithmetic is ω-consistent it may be provable that A(0), A(1), ... A(n), ... but still there may be some natural number for which the proposition fails. By Gödel's First Incompleteness Theorem, the ω-consistency of arithmetic cannot be proved, so there is an existential doubt naturally holding over every proof by the principle of mathematical induction.[316]

However, here is my attack on the principle of mathematical induction. The basic idea is to give a simple proof of a proposition which parallels standard inductive proofs, but which runs into conflict with "big cock theorems" of more advanced mathematics, in particular, Gödel's Second Incompleteness Theorem, that the consistency of Peano arithmetic cannot be proved in the system itself. Since there can be no choice between the principles from different areas of mathematics, the principle of mathematical induction has to be held as being inconsistent.

Proof: consider proposition P(n) such that a natural number n is not equal to 0 for n > 0, i.e. ~ (n=0, unless n=0), where "~" means "not."

For P(1), this says that 1 is not equal to 0, as 1 > 0. Now using the standard reasoning mathematicians use, that is true by inspection, or if one likes, by the Peano axioms for arithmetic, as 1 is the successor

---

316 http://mathworld.wolfram.com/GoedelsIncompletenessTheorem.html.

of o. Sure, logically circular, but no more so than all mathematical reasoning.

P(k) is then assumed to be true, that is, that for all k, ~(k=o), for k > o.

Then consider proposition P(k+1), that proposition k+1, is not equal to o. Yet this follows from P(k), as if it is not the case that k = o, then from P(1), P(k+1) is not equal to o either. Hence, the proposition is true by the principle of mathematical induction. It follows then that ~ (1 = o).

Yet, the proof is logically circular, as many such proofs are. Worse, the conclusion alleges the absolute consistency of arithmetic, ~ (1 = o), which is in direct conflict with the big dick theorem of Gödel, that the consistency of arithmetic is not provable in arithmetic. Hence, a contradiction.

I hope that readers are enjoying my articles as much as I dislike writing them. My goal is the noble one of dipping the academically sacred into the profanity of liquefied shit.

XVII. FUCK INFINITY:

Arguing for Strict Finitism

One of the best now arguments for the existence, or alleged existence of infinity in mathematics, outside of set theory, comes from number theory. While infinity is used in the theory of limits and hence the calculus, that symbol is one of convenience, and can be eliminated if necessary, in terms such as "increases without bonds," what the ancient Greek philosophers would call the "potential infinite," by contrast to the actual infinite. Strict finitists would say that there are only a finite number of say natural numbers, 0, 1, 2, 3, ... n ... such as made by finitists in the following papers.[317]

The conventional Platonistic mathematician will argue that given a hypothetical "last" number L, simply consider L + 1, to obtain a larger number. The question-begging assumption is made that L + 1 actually denotes, but we ignore that. Further, as shown above, even that assumption is false for my hypothetical last real number 99999....999, where adding 1 makes a smaller number!

The argument assumes that numbers exist in some non-material realm of abstract essence, as has been done from Plato to Gödel. That view is subject to metaphysical difficulties that mathematicians seldom consider, such as how is mathematical knowledge possible if the entities in question have primarily a non-material existence,

---

[317] https://www.math.unihamburg.de/home/loewe/HiPhI/Slides/bendegem.pdf; http://www.jstor.org/stable/; https://projecteuclid.org/download/pdf_1/euclid. ndjfl/1093634481; https://www.jstor.org/stable/pdf/2273760.pdf?refreqid=excelsior %3A5a69fb898236f1f5f11fb5b134fa9c78.

since causal interaction with the brain is ruled out.[318]

If a Platonist account of mathematical entities is rejected, and there are many good reasons for doing so,[319] mathematics may be more closely related to human social practices. Numbers may not be infinite, nor could chains of reasoning processes, since the human brain is finite, and of finite information processing capacity. Further, there will be physical limits of the possible representation of a number. A number written on a hypothetical medium, at the smallest possible length permitted by quantum mechanics, expanding the entire length of the known physical universe (assumed to be finite, about 93 billion light years), would not be computable in principle.

Suppose that the number is composed of the digits of the decimal expansion of a series of 99999.... and after each digit, the raised to a power symbol "^" is added. We can then suppose that the number is not merely written in a linear fashion on our hypothetical quantum mechanical medium, but extends in 3D space, or if string theory is correct, into 26-dimensional space (or 10 or 11). Thus, the entire universe is full up with one step ladder number. And, it is not possible to physically or conceptually add 1 to this number, because there is no space in the universe to even hypothetically add any more symbols. We could say that this number, Ʊ, is the largest natural number, even if there could be a Platonist conception of a larger number, because no bigger number could be represented. Likewise, a step ladder number comprising the entire universe could be constructed to refute the standard proof that there is an infinity of prime numbers. Note that the largest prime number found to date is:

$2^{77,232,917}$-1 with 23,249,425 digits by Pace, Woltman, Kurowski, Blosser & GIMPS (26 Dec 2017).

By our conception, that is a very small number. My dick is much bigger.

[318] http://www.columbia.edu/~jc4345/benacerraf%20with%20bib.pdf.

[319] https://www.calstatela.edu/sites/default/files/groups/Department%20of%20 Philosophy/realism_and_anti-realism_in_mathematics.pdf.

A natural number is defined to be prime if the only divisors of that number are 1 and the number itself, so that the first few primes are 2, 3, 5, 7, 11, 13, 17, 19 … Note that an odd number is not necessarily prime, 9 being an example. The even number 2 is prime, but that is the only one, because bigger numbers will be divisible by 2. There are good reasons why the number 1 is not defined as prime, or composite (not prime), not fitting the definition of either concept.[320]

Euclid (died 285 BC), in his *Elements*, has been attributed the proof that there are an infinite number of primes. Actually, Euclid in book IX, proposition 20, argued this:[321]

Proposition 20

Prime numbers are more than any assigned multitude of prime numbers.

Let *A*, *B*, and *C* be the assigned prime numbers.

I say that there are more prime numbers than *A*, *B*, and *C*.

Take the least number *DE* measured by *A*, *B*, and *C*. Add the unit *DF* to *DE*.

Then *EF* is either prime or not.

First, let it be prime. Then the prime numbers *A*, *B*, *C*, and *EF* have been found which are more than *A*, *B*, and *C*.

VII.31

Next, let *EF* not be prime. Therefore it is measured by some prime number. Let it be measured by the prime number *G*.

I say that *G* is not the same with any of the numbers *A*, *B*, and *C*.

If possible, let it be so. Now *A*, *B*, and *C* measure *DE*, therefore *G* also measures *DE*. But it also measures *EF*. Therefore *G*, being a number, measures the remainder, the unit *DF*, which is absurd.

Therefore *G* is not the same with any one of the numbers *A*,

---

[320]   https://www.quora.com/Why-is-1-neither-prime-nor-composite; http://mathforum.org/library/drmath/view/57036.html; https://primes.utm.edu/notes/faq/one.html;https://cs.uwaterloo.ca/journals/JIS/VOL15/Caldwell1/cald5.html.

[321]   https://mathcs.clarku.edu/~djoyce/java/elements/bookIX/propIX20.html.

B, and *C*. And by hypothesis it is prime. Therefore the prime numbers *A*, *B*, *C*, and *G* have been found which are more than the assigned multitude of *A*, *B*, and *C*.

Therefore, *prime numbers are more than any assigned multitude of prime numbers*.

At no point did Euclid use the notion of infinity, which the Greek mathematicians were generally opposed to, following Aristotle in holding only to the "potential infinite."[322]

A very good article discussing Euclid's proof and some flawed interpretations of it is by M. Hardy and C. Woodgold, "Prime Simplicity," *The Mathematical Intelligencer*, (2009).

Prime numbers being more than any assigned multitude of prime numbers means that there is no greatest prime. As we will see, it will require further argument to justify the claim that there is a literal infinity of primes. Further, Euclid did not present the standard proof by contradiction or *reductio*, but showed that given any list of primes, one can show that a new prime can be added to the list.

That does not show, as all the textbooks state, that there is an infinity of primes. What is established is the weaker thesis that there is no largest prime. But, I doubt whether, if Platonism is rejected, that the proof which requires listing the finite list of primes could even get started if we consider that we are dealing with super-large, but still finite numbers like ʊ, where there is no way (except in the Platonistic imagination) of representing ʊ + 1.

It is of interest that one leading philosopher of logic, Wittgenstein, had doubts about the alleged infinity of the primes: T. Lampert, "Wittgenstein on the Infinity of Primes," *History and Philosophy of Logic*, vol. 29, 2008, pp. 63-81. Among other things, critiquing Euler's alleged proof of the infinity of primes, Wittgenstein said: "Euler's proof is immediately in error, as soon as prime numbers are written down in the form, $p_1, p_2, \ldots p_n$. For, if the index n is to mean an arbitrary number, then this already presupposes a law of progression, and this law can be given only in terms of an induction.

---

[322]   https://link.springer.com/chapter/10.1007/978-1-4899-0007-4_4;   http://www.math. tamu.edu/~dallen/masters/infinity/content2.htm.

Thus, the proof presupposes what it is supposed to prove." (p.74) In other words, the use of the variable "n," which can take any natural number as its value, thus smuggles in the idea of infinity, that an infinite number of primes exist. Hence, the standard proofs beg the question, committing the fallacy of *petitio principii*, and are therefore not justified.

Regardless of this, even on Platonistic assumptions, the best that the Euclidian argument shows is that there is no largest prime, and it does not follow from this that there is an infinity of primes. Indeed, set theory does not prove that there is an infinite set of objects, but adds an axiom of infinity to get the desired result, something recognised by Zermelo in 1908. It was also added as an axiom by Russell and Whitehead in *Principia Mathematica*. However, it was pointed out by F. Ramsey, "The Foundations of Mathematics," *Proc. London Math Soc*, vol. 25, 1926, pp. 338-384, that the axiom of infinity needs to be taken as a primitive and cannot be proven. I will argue in other papers, that it can be refuted though.

The critical analysis continues in "Counting Cantor: A Refutation of the Power Set Axiom."

XVIII. COUNTING CANTOR:

A Refutation of the Power Set Axiom

For those championing the concept of the actual infinite in mathematics, and physics, set theory comes to their rescue. In this paper, I will refute a major theorem, or alleged theorem, purporting to show that there are different levels of infinity regarding sets, Cantor's power set theorem.

Wilfred Hodges, "An Editor Recalls Some Hopeless Papers, *Bulletin of Symbolic Logic*, vol. 4, no 1, 1998, pp. 1-16, says while gloating about cranks that attempted to refute Cantor, "None of the authors showed any knowledge of Cantor's theorem about the cardinalities of power sets." (p.2) So, let's examine this.

Cantor's power, or the axiom of powers, states that for any set S, there exists a collection of sets P_{ower} (S), which contain in its elements, all of the subsets of the given set S. Thus, for example, if S = { a, b }, then P_{ower} (S) = { ϕ, {a}, {b}, {a, b}}, where "ϕ" is the empty set. As the sets get bigger, so does the power set. In fact:

"If S is a finite set with $|S| = n$ elements, then the number of subsets of S is $|P(S)| = 2^n$. This fact, which is the motivation for the notation 2^S, may be demonstrated simply as follows,

First, order the elements of S in any manner. We write any subset of S in the format {γ_1, γ_2, ..., γ_n} where γ_i, $1 \leq i \leq n$, can take the value of 0 or 1. If $\gamma_i = 1$, the i-th element of S is in the subset; otherwise, the i-th element is not in the subset. Clearly the number of distinct subsets that can be constructed this way is 2^n as $\gamma_i \in \{0, 1\}$."

Cantor's diagonal argument shows that the power set of a set (whether infinite or not) always has strictly higher cardinality than the set itself (informally the power set must be larger than the original set). In particular, Cantor's theorem shows that the power set of a countably infinite set is uncountably infinite. The power set of the set of natural numbers can be put in a one-to-one correspondence with the set of real numbers.[323]

That is a fundamental argument for the different infinities of sets. The power set axiom is found in many pedigrees of set theory, especially Zermelo-Fraenkel set theory,[324] but not in all, such as Kripke-Platek set theory.[325]

Cantor himself realised that a paradox arose from considering the set S, the set of all sets, which would contain the power set, but the power set would also have a higher cardinality than it, or precisely there is no greatest cardinal number, hence the set of infinite sizes is itself greater in cardinality than its power set. The difficulty has been handled by redefining the concept of set, so that collections of this size became proper classes.

Nevertheless, while in the 20th century logicians dealt with paradoxes arising from considering sets, there are also more challenging paradoxes involving the elements of sets, or not sets of sets, but sets of infinite elements that are not sets, or even mathematical objects at all.

For example, Patrick Grim, "There is No Set of all Truths," *Analysis*, vol. 44, 1984, pp. 206-208 (also P. Grim, *The Incomplete Universe: Totality, Knowledge, and Truth*, (MIT Press, Cambridge, MA, 1991)), showed that the set of truths was in conflict with Cantor's power set theorem. And, we can add the set of facts. The set of facts has the power set P_{ower} (F), but for each set of facts F in that power set, there will be another fact, such that F contains those

---

[323]   https://en.wikipedia.org/wiki/Power_set.

[324]   https://en.wikipedia.org/wiki/Zermelo–Fraenkel_set_theory.

[325]   https://en.wikipedia.org/wiki/Kripke–Platek_set_theory;https://arxiv.org/abs/1801.01897.

facts that it does. Hence, by Cantor's power set theorem, there will be more facts in F than in its power set, a contradiction. Everyone, to date has concluded that these perfectly intuitively meaningful sets do not exist. But, if this is so, then set theory itself is flawed because a general mathematical theory should not generate paradoxes when applied to objects of the world, or thought, like truths and facts: it defeats the very purpose of a scientific theory. The same paradox comes to exist with other items that could serve as elements of a set, such as the set of all mathematical objects, the set of abstract entities, the set of all items that can be (consistently) contemplated, and so on.

Grim was concerned with using this style of argument against the existence of an omniscient being, who knew all facts/truths. We can leave that to one side. More recently he has given some indication that these paradoxes indicate that not all is well with received set theory itself, as detailed in N. Rescher and P. Grim, *Beyond Sets: A Venture in Collection-Theoretic Revisionism*, (Ontos, Verlag, 2011). Rescher and Grim state:

> Set theory was born in paradox, was shaped by paradox, and continues to carry the threat of paradox into its current adolescence. Properly understood ... the threat of contradiction is not merely formal and is not to be evaded by merely formal techniques. The fact that there can be no set of all non-self-membered sets might be shrugged aside as a minor logical surprise. Beyond Russell's paradoxical set, however, there lies the serious philosophical difficulties of coherently conceptualising a set of all things, the realm of unrestricted quantification (or even the sense of restricted quantification), the totality of all events, all facts, all propositions, or all that is true. Sets are structurally incapable of handling any of these." (p.6)

That is an excellent argument for the rejection of set theory, even in mathematics, since there are clear counter-examples to fundamental axioms, such as the power set axiom. And, at least, the rejection of the power set axiom, via these counterexamples, would severely cripple the Cantorian project of the infinite, that of countable and uncountable sets.

Take that Cantor, you piece of transfinite shit!

XIX. FUCK INFINITY: PART I

After watching *Avengers: Infinity War* and *Endgame* it is a good time to attack the very idea of infinity. This is what I do here. I am grateful to Professor James LaFond for publishing this paper where I open my cloaca and let fly at the mathematicians and physicists, whom I have a pathological hatred for. Stuck up geek cock-suckers, I say. And, in a sense, logic can be a martial art too, so without further masturbation, let's get down to it.

THE PROBLEM OF INFINITY

In a very interesting paper by Amanda Gefter, "Infinity's End: Time to Ditch the Never-Ending Story?" *New Scientist*, August 14, 2013, we are given a concise diagnosis of one of the many things, I believe, wrong with mathematical physics. Infinities abound in mathematics—in set theory, analysis, geometry and the calculus— but infinities cause great problems in physics. As she says:

> Trouble is, once unleashed these infinities are wild, unruly beasts. They blow up the equations with which physicists attempt to explain nature's fundamentals. They obstruct a unified view of the forces that shape the cosmos. Worse of all, add infinities to the explosive mixture that made up the infant universe and they prevent us from making any scientific predictions at all.

Infinities arose in the early study of the electron, but these problems were dealt with, certainly by the time quantum mechanics was worked out. However, further infinities arose in quantum electrodynamics, the quantum mechanical theory of electromagnetic forces. This was in turn dealt with by "renormalization," a kind of way of "dividing" out the infinities, which some philosophers of science (e.g. P. K. Feyerabend, gee, we don't read much about him now do we?), believed was problematic. Feyerabend said in *Against Method* (1975), p. 61:

> This procedure consists in crossing out the results of certain calculations and replacing them by a description of what is actually observed. Thus, one admits, implicitly, that the theory is in trouble while formulating it in a manner suggesting that a new principle has been discovered.

So, the method is at best, less than rigorous; at worse, invalid, if not delusional.

The infinities arising in the general theory of relativity are not so easily dealt with. Conditions inside a black hole seem to involve an infinite density of matter and an infinite warping of space-time, whatever that means. If the matter inside a black hole is infinitely dense, then it is reasonable to suppose that the black hole itself is infinitely dense, and if this is so, the gravitational attraction between the black hole and the rest of the universe should also be "infinite." Matter should disappear into the black hole like a giant vacuum cleaner. As black holes, if they exist, are not that powerful, then it is unlikely that matter inside is really "infinitely" dense.

Gefter points out that in the present cosmological theory of inflation, in the first instances after the Big Bang, there was a rapid expansion of matter and space-time. This, however, leads to continuous universe creation, with the inflation of more space-time and the creation of multiple universes. Apparently in these multiverses, anything can happen. Could I sell a Kindle book? No, anything but that is possible. Some of these universes will have radically different laws of nature. All this arises from the assumption that space-time is like the real number line, a continuum infinitely divisible; without that assumption there will be no infinite explosion of universes.

INFINITY IN MATHEMATICS

Since Aristotle, probably most mathematicians, have rejected the use of an actual infinity, and instead accepted the idea of a potential infinity, that one could add 1 to any natural number (to take the example of arithmetic), without reaching a last number. However, all that changes with Georg Cantor (1845-1918) and his work in transfinite set theory.

Galileo noted in 1638, that the set of all natural numbers {1, 2, 3, 4,...}, could be put in a 1-1 correspondence with one of its subsets {1, 3, 5, 7, ...}, the set of odd natural numbers, even though the set of natural numbers has numbers that are not in that subset. Cantor did not see this as a paradox and said that the sets are the same size, both denumerably or countably infinite. However, by an argument known as the "diagonal argument," he allegedly showed that the set of natural numbers was a smaller infinite set than the set of real numbers, because a parallel type of 1-1 correspondence could not be set up. In a hypothetical list, it is allegedly possible to always change a number in the diagonal of the set of reals, to create a real not in 1-1 correspondence with the natural numbers. Hence, the real numbers were of a higher cardinality than the natural numbers, nondenumerably or uncountably infinite, and that is just the start of the transfinite stairway to Platonic heaven. More on this bs below.

PARADOXES: LOGICAL

Not all mathematicians were pleased with the arrival of the transfinite enfant terrible. The great French mathematician Henri Poincaré (1854-1912) regarded Cantor's transfinite set theory as a "disease from which one has recovered." But he got that one wrong.

Around the turn of the 20th century a series of logical paradoxes in set theory, such as Russell's paradox, of the set of all sets not members of themselves (a set which is a member of itself, if and only if it is not a member of itself and which is therefore contradictory), rocked the mathematical world. Actually, just a few elites in a closed circle got "rocked;" nobody else cared a fuck. Anyway, infinity was thought to have had a part to play in this, but it was not the only problem; self-reference seemed problematic as well. Nevertheless, set theory paradoxes continued to be uncovered, such as the paradox of the set of all truths, which conflicts with one of old Cantor's central theorems, the power set theorem. (See P. Grim, *The Incomplete Universe*, (1991), pp. 92-93.) Also, more above on this.

Perhaps what really did cause irritable bowel syndrome among the logicians, was that it was proven on the basis of the axioms of standard set theory that it was not possible to prove or disprove various transfinite set theoretical statements, such as whether or not there were other transfinite sets between the known ones, or even the size of certain large transfinite sets (the continuum hypothesis). It is somewhat ironic to note that one of the mathematic establishment's leading symbolic logicians, Abraham Robinson, who himself developed a theory of infinitesimals (known as non-standard analysis), said in his 1973 retiring presidential address at the Annual Meeting of the Association for Symbolic Logic ("Metamathematical Problems," *Journal of Symbolic Logic*, vol. 38, 1973, pp. 500-516) the following: "While others are still trying to buttress the shaky edifice of set theory, the cracks that have opened up in it have strengthened my disbelief in the reality, categoricity or objectivity, not only of set theory but also of all other infinite mathematical structures, including arithmetic." (p. 514) I wonder if his skepticism extended to his own non-standard analysis? It should have.

Robinson said as well: "In terms of the foundation of mathematics, my position (point of view) is based on the following two main principles (or opinions); (1) No matter which semantics is applied, infinite sets do not exist (both in practice and in theory). More precisely, any description about infinite sets is simply meaningless. (2) However, we still need to conduct mathematical research as we have used to. That is, in our work, we should still treat infinite sets as if they realistically exist." A. Robinson, *Formalism 64, Logic, Methodology, and Philosophy of Science*, in Y. Bar-Hillel (ed), "Proceedings of the 1964 International Congress," (North Holland,

1964). But if the concept of infinite sets is really problematic, why should one therefore have confidence in their use even on pragmatic grounds? Why, therefore, should the concept of infinite sets have utility, any more than say, fairies at the bottom of the garden? Why should one trust mathematical reasoning employing the concept of an infinite set?

As an example of some of the problems that infinity raises for conventional mathematics, consider how use of the infinite raises immediate problems for classic probability theory. The probability of choosing any particular real number in the range [0,1] is zero, but some number would be chosen, so the probability, understood intuitively as the chance of a draw, is non-zero. A variant: consider a circle with a non-denumerable infinity of points. A "magic" rotating pointer is assumed to be able to stop at any point. The probability of any point being stopped at is zero, as the denominator in the probability ratio is "infinite," yet the pointer is assumed to be able to be capable of stopping at any point. A final example. How probable is an infinite sequence of heads in an imaginary infinite number of tosses? (See *Analysis*, vol. 67, no. 3, 2007, pp. 173-180.) One argument is that the probability of heads of such a toss is ½.½. ½ ..., which converges to zero. But, by another argument the set {h, h, h,...} is one possible sample space, and so is logically possible, and thus must have a non-zero probability. There are many attempts to get around these particular problems, but at this stage in the argument, the examples illustrate the problems which infinity raises.

PARADOXES: PHYSICAL – ZENO'S PARADOXES

Paradoxes also arise when attempting to model infinity experiments in the physical world, especially involving so-called "super-tasks." One problem, given by William Lane Craig ("A Swift and Simple Refutation of the Kalam Cosmological Argument?" *Religious Studies*, vol. 35, 1999, pp. 57-72), concerns someone who has an infinite number of marbles, and who wants to give you an infinite number of marbles too. Thus, all of the marbles M, could be given, so M-M=0.

Or only the odd numbered marbles could be given, so M-M=M. Hence dividing and subtracting equal amount gives contradictory results. Quick as a flash, the mathematician will counter that Cantorian transfinite "arithmetic" is not like ordinary arithmetic, permitting subtraction and division, and of course, they will be right. Indeed, Craig's argument fails even for the extended reals, with infinity minus infinity (∞ - ∞) being undefined. Nevertheless, this thought experiment does not appear to be logically contradictory (it begs the question to assume the correctness of the Cantorian position, when it is that position which is open to critique), and combined with other problems, begins to build up a balance of reason against Cantorianism. See Kip Sewell, *The Case Against Infinity*, (2010).[326]

Another related area where problems of infinity directly come into play with physical reality are Zeno's paradoxes, posed by Zeno of Elea (born C. 490 BC). A standard introduction is "Zeno's Paradoxes," *Stanford Encyclopedia of Philosophy*.[327]

Zeno believed that motion did not exist because he was a cosmic monist who followed his teacher Parmenides in holding that all that existed was an undivided "one." He produced a number of ingenious paradoxes dealing with space, time, motion, change and infinity, that are still being refuted today, over and over again. All on a different basis.

The *Dichotomy Paradox*: for a body to reach some given point, it must first travel half of that distance. But, before it can complete that it must travel half of the previous distance, and so on *ad infinitum*, so that it can never get started, having to pass through an infinite number of such divisions.

Achilles and the tortoise: there is a hypothetical race between Achilles and a tortoise, with the tortoise being given a lead. Achilles must first travel the distance from the starting point to the tortoise, but then the tortoise has advanced further, and that distance must then be covered and so on, so Achilles never catches the tortoise.

[326] http://philpapers.org/archive/SEWTCA.

[327] http://plato.stanford.edu/entries/paradox-zeno/.

The *Arrow Paradox*: the orthodox account of motion (or velocity, if directed), has it that motion is a relational state. Motion/velocity is the occupation of different places at different times, so there is no instantaneous state of motion. There is no real difference between a body in motion and a body at rest at an instant, because both are at that instant located at a single point in space. However, this leads directly to *Zeno's Arrow Paradox*. If one considers an arrow, the point of the arrow at instant t_0 is not advancing, being at a point A in space. But, to move between points A and B, the arrow needs to progress through all of the points between A and B, and if it does not progress on its journey, at any of these points, then it cannot traverse the line segment AB. Hence, it seems that the arrow cannot move – but I bet that Zeno wouldn't have stood down range from one!

Another related problem, which was not given by Zeno, is the instant of change problem. Before a time t_0 a system was in state S_0, but after t_0 the system is in state S_1. So what state was the system in at t_0? The possibilities are: (1) in one of S_0 or S_1; (2) in neither S_0 nor S_1 or (3) in both S_0 and S_1. Applied to motion the states are "at rest" and "in motion" and none of the options seem to be justified. See J. R. McKie, "Transition and Contradiction," *Philosophica*, vol. 50, 1992, pp. 19-32. So, what the fuck is going on?

Now, many think that the differential calculus does give us an instantaneous velocity, that is a velocity at an instant. But, according to the limit definition, a limit of velocities is considered at successively shorter periods of time, converging to a given instant. That, however, assumes that motion is already occurring, and so begs the question against Zeno. See J. Lear, "A Note on Zeno's Arrow," *Phronesis*, vol. 26, 1981, pp. 91-104.

Another paper says that in the Arrow Paradox Zeno assumed that the velocity v=0, but what we have is an indeterminate form of the type zero over zero i.e. v=0/0, which could be any real number. See M. Zangari, *Australasian Journal of Philosophy*, vol. 72, 1994, pp. 187-204. To my skeptical mind, this still does not say what v actually is, so Zeno's difficulties remain.

W. McLaughlin, "Resolving Zeno's Paradoxes," *Scientific American*, no.5, 1994, argued that the solution lies in supposing that motion occurs in infinitesimals, non-standard numbers, greater than zero, but less than any real number. Between real instants, there are allegedly infinitesimals, and objects move through them. But, that seems to make matters even worse, because now one also has to explain how objects move through infinitesimals to real points.

Some very good papers outlining the difficulties with the mathematical physics approach to Zeno's paradoxes are: A. Papa-Grimaldi, "Why Mathematical Solutions to Zeno's Paradoxes Miss the Point: Zeno's One and Many Relation and Parmenides' Prohibition," *Review of Metaphysics*, vol. 50, 1996, pp. 299-314; A. Papa-Grimaldi, "The Presumption of Movement," *Axiomathes*, vol. 17, 2007, pp. 137-154; T. Glazebrook, "Zeno Against Mathematical Physics," *Journal of the History of Ideas*, vol. 62, 2001, pp. 193-210.

The answer lies in the idea that space and time are discrete, not continuous, that is, that there are both quanta of space and time. This idea raises some problems of working out a finite geometry. But the McKie paper cited earlier discusses this and makes some useful suggestions, as do these two papers; J. P. van Bendegem, "Zeno's Paradoxes and the Tile Argument," *Philosophy of Science*, vol. 54, 1987, pp. 295-302 and P. Forrest, "Is Space-Time Discrete or Continuous? – An Empirical Question," *Synthese*, vol. 103, 1995, pp. 327-354.

Discussions of Zeno's paradoxes have also considered whether an infinite number of tasks can be completed in a finite time (a "supertask"), and whether or not infinite processes can be completed in reality. J. Benardete, *Infinity*, (Oxford University Press, 1964), p. 237, asked us to consider a book which is in one-inch-thick, with the first page of the book being ½ inches thick, the second page ¼ inches thick, and so on *ad infinitum*. While the whole book is one-inch-thick (as the series converges), if one looks at the book from the back cover there is nothing to see as there is no last page in the book. As a form of Zeno's paradox, any book, or object could be modeled in this way.

John Earman and John Norton in their paper, "Infinite Pains: The Trouble with Supertasks," in *Benacerraf and His Critics* (1996), see no problem with an infinite series of tasks being completed in a finite time. Their example is an idealized perfectly elastic ball bouncing on a hard surface. With a single release, with each bounce the speed on rebound is reduced by a fraction of its speed prior to the bounce. Assuming that each bounce takes no time at all, it can allegedly be shown that the time between successive bounces forms a converging series, so an infinite number of bounces can be completed in a finite time. But, I doubt it. Even for an ideal physical system, the assumption that a bounce takes no time at all is incoherent, because a "bounce" is a physical event that by definition takes time, however small. Hence, the thought experiment is flawed. And, where is the proof that the ball actually stops? It may look like it has, but, who knows?

The "Thomson lamp" was devised as a thought experiment to show the incoherence of the idea of supertask. The hypothetical lamp, which doesn't burn out etc., has an on/off switch, turned on in the 1st minute, off in the next half minute, on in the next ¼ minute, and so on. At the end of two minutes is the Thomson lamp on or off? The series 1-1+1-1... diverges, so no meaningful answer is possible, unless we say that the lamp is both on and off, which paraconsistent logicians may go for. The generally accepted rebuttal is by P. Benacerraf, that being on or off only applies to times prior to the supertask being completed, and that in itself shows nothing about the lamp's state at 2 minutes. He should have been called out on this. If that was so, then there is nothing at all to justify any conclusion about the state of the lamp at 2 minutes. That seems to be even more problematic.

Earman and Norton say that "informally we assume that if a lamp is left unswitched, it persists in its current state. Therefore, the state of the lamp at a time when it is not switched is automatically fixed by the prior history of switching." (p.238) They claim that if the "persistence assumption" is rejected and some other way of fixing the state of the lamp at 2 minutes is presented, then there is no logical contradiction. But, they don't give any such method, and it is irrelevant any way because the state of the hypothetical lamp, by hypothesis is fixed by its prior states. That is just how the problem is defined. See C. Ray, *Time, Space and Philosophy* (1991).

One way out of all of this is to move to a finite mathematics. There is a considerable amount of work in this genre and some papers include: J. P. van Bendegem, "Alternative Mathematics: The Vague Way," *Synthese*, vol. 125, 2000, pp. 19-31; J. P. van Bendegem, "Finite, Empirical Mathematics: Outline of a Model," (1987); D. H. Sanford, "Infinity and Vagueness," *Philosophical Review*, vol. 84, 1975, pp. 520-535; P. T. Shepard, "A Finite Arithmetic," *Journal of Symbolic Logic*, vol. 38, 1973, pp. 232-248; J. Mycielski, "Analysis Without Actual Infinity," *Journal of Symbolic Logic*, vol. 46, 1981, pp. 625-633; S. Lavine, "Finite Mathematics," *Synthese*, vol. 103, 1995, pp. 389-420; S. Lavine, "Understanding the Infinite," (1994); J. P. van Bendegem, "Classical Arithmetic is Quite Unnatural," *Logic and Logical Philosophy*, vol. 11, 2003, pp. 231-249.

Hence, there are finitist alternatives, but this does not in itself refute infinitism, a task which still needs to be undertaken. Let us first discuss some aspects of physics where infinity seems to have made a home, then move to mathematics.

XX. FUCK INFINITY: PART II

Fairytale Physics

There are two areas of physics where infinity seems to have taken hold like tapeworms; string theory and black holes. A good book critiquing string theory is Jim Baggott, *Farewell to Reality; How Fairytale Physics Betrays the Search for Scientific Truth*, (Constable, London, 2013). I will have something to say about one aspect of the mathematics of string theory in another paper, perhaps at this distinguished site, but here I will restrain my hatred to just focus on one question: what are these strings which are supposed to unify physics?

What exists, the stringers propose, are loops of cosmic "string," vibrational patterns. But, patterns of what? Originally these strings vibrated in 26 space-time dimensions (25 dimensions of space, 1 time), but this was reduced to 10 dimensions (9 space, 1 time). Make up your fucking mind. However, in M-theory this was back up to 11 dimensions. M-theory is supposed to explain the universe, but nobody really knows what it is. Sounds like bullshit to me.

The strings are not dimensionless, as the point particles were in quantum field theory. As Brian Greene puts it in *The Elegant Universe*, (Norton, New York, 1999), "the spatially extended nature of a string is the crucial new element allowing for a single harmonious framework incorporating both (gravitational and quantum mechanical) theories." (p. 136) However, as Steven Rosen, "Quantum Gravity and Phenomenological Philosophy," *Foundation of Physics*, vol. 38, 2008, pp. 556-582, argues, the idea of fundamentality, with a finite extension, does imply cuttability and division, which means that string theory is metaphysically contradictory.

We do not know then what makes up the string, or even why it vibrates. Baggott concludes that this physics is "informed not by the practical necessities of empirical reality, but by imagination constrained only by the internal rules of an esoteric mathematics and an often rather vague connection with problems that theoretical physics beyond the standard model is supposed to be addressing. No amount of window dressing can hide the simple fact that this is all *metaphysics*, not physics." (p. 202)

The other area of physics requiring criticism is the theory of black holes. Popular science books, television shows and films abound with stories about black holes – gravity collapsed stars – which according to some accounts are bridges to other universes, or some such nonsense. The alleged existence of black holes is taken to confirm the genius of physicist Albert Einstein, as black holes are one of the theoretical consequences of the General theory of Relativity, the theory of gravity. But is this narrative actually true?

Taiwanese cosmologist and physicist, Wun-Yi Shu, "Cosmological Models with No Big Bang," arXiv:1007.1750v1, via:ThePhysicsArXivBlog, has presented a cosmological model with no Big Bang, no beginning and no end of the universe. Time and space can be converted into one another with the conversion factor being a varying speed of light, depending on a varying gravitational constant. As the universe expands, time is converted into space and mass is converted into length, and the opposite when the universe contracts. "We view the speed of light as simply a conversion factor between time and space in space time. It is simply one of the properties of space time geometry. Since the universe is expanding, we speculate that the conversion factor somehow varies in accordance with the evolution of the universe, hence the speed of light varies with cosmic time." Take that Uncle Albert! Although not explicitly directed against the existence of black holes, the hypothesis of mass being converted into length as the universe expands could be taken as a challenge to the permanent of black holes.

In early 2014, black hole father Stephen Hawking, who is now dead, released an on-line paper, "Information Preservation and

Weather Forecasting for Black Holes." He had come to reject the existence of conventional black holes, arguing that they were merely "grey;" gravity collapsed stars without event horizons. The event horizon was supposed to stop light from escaping. This change in position was Hawking's response to the so-called "information paradox." My, how things change. He has changed. And, now, sadly, the great man is dead. Such is this cruel world when in the end, rot is all.

The information loss paradox concerns what happens to "information" that goes into a black hole. The heart of the problem is reconciling quantum mechanics, the theory of the small, with general relativity, the theory of the big (e.g. cosmological structures such as galaxies and the universe itself). General relativity implies that black holes can exist, being a singularity, which is essentially infinite, constituting a "hole" in space-time. Beyond the event horizon, the point of no return, no matter-energy or information escapes. But, this conflicts with quantum mechanics' law of conservation of information, and ultimately the law of conservation of mass-energy, of which information is a representation.

In 2014 cosmologist Laura Mersini-Houghton argued that black holes do not exist. This was done in two internet papers: (1) L. Mersini-Houghton, "Backreaction to Hawking Radiation on a Gravitationally Collapsing Star I," arXiv:1406 1525v1[hep-th], June 5, 2014; (2) L. Mersini-Houghton and H. P. Pfeiffer, "Back-Reaction of the Hawking Radiation Flux on a Gravitationally Collapsing Star II: Fireworks Instead of Firewalls," arXiv: 1409.1837v1[hep-th], September 5, 2014.

Without the mathematics, the basic critique of black holes is that the Hawking radiation emitted by the star during its gravitational collapse slows down the collapse of the star and substantially reduces its mass. The collapsing star "bounces" before reaching the event horizon, with the area radius increasing. The star ceases to collapse at a radius, which is greater than the event horizon would be, and the core explodes. But why does the star explode? Here is what the authors say: "Physically the backreaction of ingoing negative energy

Hawking radiation reduces the gravitational binding energy in the star with the maximum loss near the last stages of collapse, while taking momentum away from the star." (p. 7, paper (2), above) Good, everybody understands that.

If all of this is correct, there are profound cosmological consequences. The standard theory of the universe is that it began with the Big Bang, where all the matter/energy that now exists was compressed into a singularity, a hyper-black hole. However, if such singularities do not exist – and violating the laws of physics is a good reason for supposing that they do not – then the Big Bang is not likely to have occurred as well. Although Mersini-Houghton mentions this consequence in a press release, this conclusion deserves a separate paper of its own.

Stephen Crothers, "Simple Proof that Black Holes Have No Basis in General Relativity," at vixra.org, e-Print archive; viXra:10405.0287, has argued that general relativity does not predict black holes. He pinpoints the fault with Einstein, who for a space-time geometry=0, and with the energy-momentum tensor=0 and material sources=0, claimed that a massive star can collapse to form a black hole. That implies that the energy-momentum tensor both includes and does not include a material source, a contradiction Crothers points out.

MATHEMATICS, INFINITY, AND CONTRADICTION

There are many contemporary critiques of the use of the actual infinite in mathematics. The most radical come from Chinese logicians led by Wujia Zhu, Department of Computer Science, Nanjing University of Aeronautics and Astronautics, Nanjing, Peoples' Republic of China. The papers were published in the English language journal *Kybernetes*: (1) Y. Lin (et al.) "Systematic Yoyo Structure in Human Thought and the Fourth Crisis in Mathematics," *Kybernetes*, vol. 37, no. 3, 2008, pp. 387-425; (2) W. Zhu (et al.) "Cauchy Theater Phenomenon in Diagonal Method and Test Principle of Finite Positional Differences," *Kybernetes*, vol. 37, no. 3/4, 2008, pp. 469-473; (3) W. Zhu (et al.), "The Inconsistency of the Natural Number System," *Kybernetes*, vol. 37, no. 3/4, 2008, pp. 482-488; (4) W. Zhu (et al.), "Modern System of Mathematics and a Pair of Hidden Contradictions in its Foundations," *Kybernetes*, vol. 37, no. 3/4, 2008, pp. 438-445; (5) W. Zhu (et al.), "Modern System of Mathematics and Special Cauchy Theater in its Theoretical Foundation," *Kybernetes*, vol. 37, no. 3, 2008, pp. 458-464; (6) W. Zhu (et al.), "Wide-Range Co-Existence of Potential and Actual Infinities in Modern Mathematics," *Kybernetes*, vol. 37, no. 3 /4, 2008, pp. 433-437; (7) W. Zhu (et al.), "Descriptive Definitions of Potential and Actual Infinities," *Kybernetes*, vol. 37, no. 3 /4, 2008, pp. 424-432; (8) W. Zhu (et al.), "Mathematical System of Potential Infinities (I) – Preparation," *Kybernetes*, vol. 37, no. 3 /4, 2008, pp. 489-493; (9) W. Zhu (et al.), "Mathematical System of Potential Infinities (II) – Formal Systems of Logical Basis," *Kybernetes*, vol. 37, no 3/4, 2008, pp. 495-504; (10) W. Zhu (et al.), "Mathematical System of Potential Infinities (III) – Metatheory of Logical Basis," *Kybernetes*, vol. 37, no. 3 /4, 2008, pp. 505-515; (11) W. Zhu (et al.), "Mathematical System of Potential Infinities (IV) – Set Theoretical Foundation," *Kybernetes*, vol. 37, no. 3 /4, 2008, pp. 516-525; (12) W. Zhu (et al.), "The Inconsistency of Countable Infinite Sets," *Kybernetes*, vol. 37, no. 3 /4, 2008, pp. 446-452; (13) W. Zhu (et al.), "Inconsistency of Uncountable Infinite Sets Under ZFC Framework," *Kybernetes*, vol. 37, no. 3 /4, 2008, pp. 453-457; (14) W. Zhu (et al.) " Intention and Structure of Actually Infinite, Rigid Sets," *Kybernetes*, vol. 37, no 3 /4, 2008, pp. 534-542; (15) W. Zhu (et al.), "Problem of Infinity Between Predicates and Infinite Sets," *Kybernetes*, vol. 37, no. 3, 2008, pp. 526-533.

What is of particular interest is an alleged proof of the inconsistency of the natural numbers, a refutation of Cantor's diagonal method and a proof of infinite sets. This material is too detailed to review here, but if it proves to be correct, it is "game over" for infinity, and probably for much conventional mathematics. So, all of you pricks on the internet attacking "cranks," here is something for you to get your claws and beaks into, you geek pieces of shit. Yeah? Well, fuck you too!

As has been said, Cantor's diagonal argument has been used to "prove" that there are allegedly different levels of infinity; namely that the set of real numbers is "uncountable" or nondenumerable, while the set of natural numbers {1, 2, 3, …}, is countable or denumerable. Thus, the cardinality of the set of real numbers is of a "higher" infinity than the set of natural numbers. Many other results can also be allegedly "proved" by the so-called *diagonalization method*, such that the set of all subsets of natural numbers is "uncountable."

The diagonalization argument involves setting up a table with the natural numbers 1,2,3, …, going down the first column. One seeks to put the natural numbers into a 1-1 correspondence with the real numbers. Thus, 1 might be paired with the real number $R_1 = d_{11} d_{12} d_{13}... d_{1k}...$, where "$d_{1k}$" are digits. Proceeding in this way, a table is set up. Cantor then sought to show that a 1-1 correspondence can be counter-exampled by constructing a new real number which does not occur in the table by starting at the left-hand corner of the array and going down the diagonal and changing each decimal by taking 1, if d_{1k} is greater than 1, and if $d_{1k}=0$, adding 1. This supposedly creates a new real number not on the list, so the 1-1 correspondence doesn't allegedly exist. If the new number was added to the top of the list and paired off with the number 0, Cantorians claim that a new application of the diagonal method would generate a new real number not on the list, and so on. See G. Hunter, *Metalogic*, (Macmillan, London, 1971), pp. 22-24.

Those rejecting Cantor's diagonal argument, on the internet, are regarded as "cranks" by the mathematical establishment, which I take as a badge of honor. However, one Cantorian has asked; "What if Cantor's Proof is Wrong?"[328] It is pointed out there that there is in fact a well-grounded mathematical reason to doubt the diagonal argument in this context, because of the Löwenheim-Skolem Theorem: if a first order theory has a model, then it has a denumerable model, if the theory is consistent. This model is one with no uncountable/nondenumerable sets. As the above mathematician puts it: "In such a model, everything we think we are saying about the real numbers

---

328   http://rjlipton.wordpress.com/2011/10/21/what-if-cantos-proof-is-wrong/.

is translated into an equally meaningful assertion about a set that is actually countable. This is ultimately because in a logical language we can *say* only countably many things before breakfast, and in a first-order language we can talk about only one thing at a time. We cannot actually say – or believe – uncountably many things before breakfast." Not a good start to the Cantorian day. This consequence of the Löwenheim-Skolem theorem, known as Skolem's paradox, is not a paradox in the sense of a contradiction, but a result which flies in the face of Cantorianism. It does not show that there are no nondenumerable sets, only that under the conditions of the theorem, such sets can be effectively eliminated, which is an equally as deadly result. See B. Slater, "Logical Paradoxes," *Internet Encyclopedia of Philosophy*.[329] It is worth noting, as Slater does in this piece, that it is arguable that Cantor's diagonal argument is actually a paradox, because a direct application of it yields Richard's paradox, which is generally accepted as a real paradox. If one cannot distinguish "good" against "bad" applications of Cantor's argument, then the whole thing will need to be rejected.

P. O. Johnson, "Wholes, Parts and Infinite Collections," *Philosophy*, vol. 67, 1992, pp. 367-379, argued against the idea that infinite sets can be placed in a 1-1 correspondence: "We cannot match terms unless we know in advance that each series or set contains the same number, and there is no other justification for saying that they 'correspond.' This is, of course, exactly the opposite argument to Cantor's. Where Cantor argues that, where there is a one-to-one correspondence between two infinite classes, both must contain the same number of objects, I say that unless some classes can be shown to contain the same number of objects, their terms cannot be said to correspond," (p. 372) It is an assumption that the notion of a 1-1 correspondence between infinite sets is meaningful, and can be established. Cantor's proof is non-constructive, and the skeptic should ask to see the "supertask" of actually producing the diagonal number completed. Go on, fucking show us! Who knows what could happen in reality as one attempts to construct such a number; maybe an evil demon will slay one, or the sky will fall?

---

[329] http://www.iep.utm.edu/par-log/.

Perhaps there is some unknown physical law that would prevent one "completing" the supertask? Who the fuck knows?

Those mathematicians and logicians critical of Cantor's diagonal argument, are usually critical of received set theory. For example, N. J. Wildberger, "Set Theory: Should You Believe?"[330] has said: "If you have an elaborate theory of 'hierarchies upon hierarchies of infinite sets, in which you cannot *even in principle* decide whether there is anything between the first and second 'infinity' on your list, *then it's time to admit that you are no longer doing mathematics.*" The same skepticism has been expressed by Wildberger in his paper, "Numbers, Infinities and Infinitesimals."[331] If forcing axioms are added to ZFC set theory, then the continuum hypothesis is false, but if the "inner-model" axiom "V=ultimate L" is added, then the continuum hypothesis is true. So, at least from the perspective of mathematical Platonism, which holds that these entities have some sort of existence, what does "God" see: is there an infinity between the smallest infinity (the set of counting numbers) and the continuum, or not? If everything depends upon what arbitrary assumptions one begins with, then at least mathematical Platonism will bite the dust.

Ludwig Wittgenstein (1889-1951), was highly critical of set theory, especially Cantor's transfinite set theory, seeing it as "utter nonsense." See V. Rodych, "Wittgenstein's Critique of Set Theory," *Southern Journal of Philosophy*, vol. 38, 2000, pp. 281-319. He rejected the concept of actual infinity, and said that there was no actual set of all natural numbers. One of his arguments against the notion of an actual infinity was this: "Let's imagine a man whose life goes back for an infinite time and who says to us: "I'm just writing down the last digit of π, and it's a 2." Every day of his life he has written down a digit, without ever having begun; he has just finished. This seems utter nonsense, and a *reductio ad absurdum* of the concept of an infinite totality." See L. Wittgenstein, *Philosophical Remarks*, (Basil Blackwell, Oxford, 1975), § 145. Well, at least applied to physical reality.

---

[330] http://web.maths.unsw.edu.au/~norman/views2.htm.

[331] http://web.maths.unsw.edu.au/~norman/papers/Ordinals.pdf.

The Australian logician H. Slater, is another critic of set theory. In "Numbers are Not Sets," *The Reasoner*, vol. 4, no. 12, December 2010, pp. 175-176, he argues that there is a grammatical confusion in taking the number zero to be the empty set, as it is the number of elements in the empty set which is zero, not the empty set itself. Along the same lines, it can be argued that the definition of the empty set, as a set containing *no* (i.e. no number) of elements, is circular, because the concept of a number is presupposed. Consequently, defining 0 as { }, will be flawed, as defining { }, will presuppose the concept of zero in specifying that this set has no number of elements. See H. Slater, *The De-Mathematisation of Logic*, (Polimetrica, Milano, 2007), p. 19, and H. Slater, "Grammar and Sets," *Australasian Journal of Philosophy*, vol. 84, 2006, pp. 59-73. Edward Nelson, *Predicative Arithmetic*, (Princeton University Press, 1986), also viewed the standard definition of a natural number as circular. On the metaphysics of sets see: Max Black, "The Elusiveness of Sets," *Review of Metaphysics*, vol. 24, 1971, pp. 614-636.

Further, if physical quantities are not treated using set theory, so that continua are not viewed as an infinite collection of points, then, "since if there is no number of points in some stuff then there is no question of whether that number is, or is not greater than some other." See H. Slater, "Aggregate Theory versus Set Theory," *Erkenntnis*, vol. 59, 2003, pp. 189-202, p. 189; H. Slater, *Against the Realisms of the Age*, (Ashgate, Aldershot, 1998), chapter 7, "Set Theory," pp. 144-156. Set theory is not an accurate account of the way we use collectives, and of the mathematics of collectives, such as groups, flocks, and so on.

In H. Slater, *Logic Reformed*, (Peter Lang, Bern, 2002), Slater rejects the idea that there is a determinate number of natural numbers, and even if two sets were allegedly put into 1-1 correspondence, and had the same "power," they may not have the same number, as they, if "infinite," may have no determinate number at all. (p. 34) He follows Aristotle in rejecting the notion of completed infinities, so that there is no number of the natural numbers or of the continuum. (pp. 35-39) By the same line of reasoning, there are no "irrational" numbers either: "if we define them not in terms of impossible

Platonic limits but merely convergent sequences of rational numbers, then we are identifying 'irrational numbers' with certain functions, since sequences are functions from the natural numbers. But the description 'number' is then strictly a misnomer, since a function is not a number, even if each of its values is one." (p. 38)

In a paper often cited by the anti-crank mathematical thought police, Wilfred Hodges, "An Editor Recalls Some Hopeless Papers," *Bulletin of Symbolic Logic*, vol. 4, 1998, pp. 1-16, he says that the main attacks against Cantor's diagonal argument attack the elementary version of the argument using the matrix representation of the sequence of decimal real numbers, but "none of the authors showed any knowledge of Cantor's theorem about the cardinalities of power sets." (p. 2) Cantor allegedly generated a hierarchy of cardinal numbers, an infinite sequence of such infinities, each one generated by taking the power set, or set of all subsets of the preceding infinite set. However, inaccessible cardinals, the so-called "higher infinite," cannot be obtained from smaller cardinals in this way: A. Kanamori, *The Higher Infinite*, (Springer, 2003)

Although there are paradoxes associated with Cantor's theorem, these are taken to be ruled out by axiomatic set theory: K. C. Klement, "Russell, His Paradoxes, and Cantor's Theorem: Part 1," *Philosophy Compass*, vol. 5, no. 1, 2010, pp. 16-28. Nevertheless, paradoxes still escape these strictures, such as Grim's paradox of the set of all truths: P. Grim, "There is No Set of All Truths," *Analysis*, vol. 44, 1984, pp. 206-208. Suppose T is the set of all truths. P(T) is the power set of T. Then for each element s_i of P(t), there is a truth t_n. So, there will be as many elements of T as P(t), contrary to Cantor's power set theorem.

There was a lengthy debate for many years about the significance of this argument for omniscience. It was noted by some logicians e.g. R. Sylvan, "Grim Tales Retold: How to Maintain Ordinary Discourse about – and Despite – Logically Embarrassing Notions and Totalities," *Logique et Analyse*, vol. 139-140, 1992, pp. 349-374; G. Priest, *Beyond the Limits of Thought*, (2002), pp. 230-232, that the set of all truths, is unobjectionable, and certainly set theory should not rule out what objects it can encompass.

As discussed above, N. Rescher and P. Grim, *Beyond Sets: A Venture in Collection-Theoretic Revisionism*, (Transaction Books, 2011), have moved to be highly critical of set theory because of its unrealistic treatment of collectivities. In particular; "Set theory was born in paradox, was shaped by paradox, and continues to carry the threat of paradox into its current adolescence." (p. 6) An outstanding philosophical difficulty is: "coherently conceptualizing a set of all things, the realm of unrestricted quantification (or even the sense of restricted quantification), the totality of all events, all facts, all propositions, or all that is true." (p. 6) And they conclude: "Sets are structurally incapable of handling any of these." (p. 6)

With the rejection of the concept of a set, we can reject as well the Cantorian worldview that goes with it.

REALLY SMALL DICKS: INFINITESIMALS

So much then for the infinitely large; there is however even more challenges facing the mathematical skeptic in taking out the infinitely small, infinitesimals. The idea of an infinitesimal quantity played a key role in the development of the calculus, but was regarded as logically inconsistent in its earlier formations. Only in the 20th century, with the use of mathematical logic and algebra, were, allegedly, these difficulties overcome for the notion of infinitesimals. But, I doubt it.

Sir Isaac Newton's differential calculus, to use Leibniz's notation, took the "dy" and "dx" in dy/dx to be the ration of infinitesimal differences. The problem facing both Newton and Leibniz's early formations was that in working out dy/dx, the infinitesimal quantities were assumed to be non-zero in the body of the algebra, or proof, but when we reached the conclusion, to conveniently get the infinitesimals to drop out of the equation, they were assumed to be zero. Critics such as Bishop George Berkeley (1685-1753) and David Hume (1711-1776), argued that this was inconsistent, making infinitesimals both zero and not-zero. See "Continuity and Infinitesimals," *Stanford Encyclopedia of Philosophy*.[332] The problem

---

[332]   http://stanford.library.usyd.edu.au/archives/fall/2008/entries/continuity/.

was taken to have been solved by the replacement of the concept of infinitesimals with that of limits, so that, informally, dy/dx is the limit of the ratio $\Delta y/\Delta x$, as Δx tends to 0. However, even the inconsistent theory of the calculus has received a modern rehabilitation by a change from classical logic to paraconsistent logic, which tolerates inconsistencies: C. Mortensen, *Inconsistent Mathematics*, (Kluwer, Dordrecht, 1995); B. Brown and G. Priest, "Chunk and Permeate: A Paraconsistent Inference Strategy; Part I: The Infinitesimal Calculus," *Journal of Philosophical Logic*, vol. 33, 2004, pp. 379-388. In later papers I plan to take the paraconsistency position apart.

Although some mathematicians, such as Georg Cantor, unsuccessfully attempted to prove that infinitesimals were contradictory (see: M. E. Moore, "A Cantorian Argument Against Infinitesimals," *Synthese*, vol. 133, 2002, pp. 305-330), the 20th century saw the mathematical rehabilitation of the idea of the infinitesimal. The best-known account is the non-standard analysis of Abraham Robinson, which will be discussed below.

There are also the "surreal numbers" of J. H. Conway, *On Numbers and Games*, (A. K. Peters, 2001); see also P. Ehrlich," The Absolute Arithmetic Continuum and the Unification of All Numbers Great and Small," *Bulletin of Symbolic Logic*, vol. 18, 2012, pp. 1-45. Thus, $1/\omega$ is an infinitesimal, where "ω" is a von Neumann ordinal. When one can just define mathematical entities into existence, what could be easier, with all the benefits of thief over honest toil? There are technical problems with the theory that will be discussed elsewhere.[333] The problem of giving a fully general account of integration is apparently not yet solved by surrealist researchers such as Kruskal, but I could be wrong on that one, having limited research opportunities from the lunatic asylum.

E. Nelson also produced infinitesimals by adding axioms to ZFC set theory: E. Nelson, "Internal Set theory: A New Approach to Nonstandard Analysis," *Bulletin of the American Mathematical Society*, vol. 83, 1977, pp. 1165-1198. This theory is also odd. Consider what J. S. Alper and M. Bridges, "Mathematical Models and

---

[333] http://mathoverflow.net/questions/29300/whats-wrong-with-the-surreals.

Zeno's Paradoxes," *Synthese*, vol. 110, 1997, pp. 143-166, say about Nelson's infinitesimals; "The nonstandard number n can never be constructed because, as Nelson proves, anything that can be explicitly constructed using classical methods is a standard object. In some mystical sense, if it were possible to figure out what n is, then it could not be that." (p. 152) To my mind, that is an excellent reason for rejecting the theory.

In *Smooth Infinitesimal Analysis* (SIA), $f'(x)$ is defined as: $f(x+\varepsilon)=f(x)+\varepsilon f'(x)$. Here, infinitesimals are nilsquare and nilpotent; they are not identical to 0, but their squares are; what the fuck! To carry this piece of intellectual masturbation off, it is necessary to reject the logical law of excluded middle, that $p\lor\sim p$, is logically true. So, x=0 or ~(x=0), for all x, is not logically true: G. Hellman, "Mathematical Pluralism: The Case of Smooth Infinitesimal Analysis" *Journal of Philosophical Logic*, vol. 35, 2006, pp. 621-651. But, don't ask what x is. If the law of excluded middle held in SIA, then it would be provable that 1=0: see J. L. Bell, "An Invitation to Smooth Infinitesimal Analysis."[334]

Along with this, theorems such as the Intermediate Value Theorem, have to be rejected. That useful theorem is: if a continuous function f in an interval [a,b] takes values f(a) and f(b) at each point of the interval [a,b], then it also takes any value between f(a) and f(b) at any point in the interval. One could turn the tables on SIA and argue that these results are grounds for its rejection, and that rejecting the law of excluded middle is ad hoc. Apart from the inconsistency, there is no independent reason for rejecting this principle. Sure, there are alternative logics, but so what? Any logical principle can be rejected, including the law of non-contradiction. If one had the ball, one could even allow absolute inconsistencies, which would enable everything to be proved (and the negation of everything), which would get the whole game over in an instant.

The most famous theory of infinitesimals is that of Abraham Robinson's non-standard analysis: *Non-Standard Analysis*, (Princeton University Press, Princeton, 1996), and found in calculus

---

[334] http://publish.uwo.ca/~jbell/invitation%20to%20SIA.pdf.

books such J. Keisler, *Foundations of Infinitesimal Calculus*, (1976). An infinitesimal ð is an element of the non-standard real numbers and is defined for a positive ð as less than r for all real r. An infinite number is greater than all real r. Two elements x and y, are infinitely close if x-y is infinitesimal. A number z is infinite if the inverse of z, 1/z, is infinitesimal. See A. H. Lightstone, "Infinitesimals," *American Mathematical Monthly*, vol. 79, 1972, pp. 242-251.

Now if all that is needed to present a mathematical idea is a definition, and maybe freedom from absolute inconsistency, then even the sky is not the limit. Thus, we could define the hyper-infinitesimal numbers to be numbers greater than 0, but smaller than any non-standard infinitesimal, and so on for hyper-hyper-… numbers. Robinson, though, attempted to show that non-standard numbers actually existed, whatever that now means.

To do this he constructed non-standard models of the reals, and made use of the model-theoretic version of the *compactness theorem*: if every finite subset of the set of proper axioms of a first order theory T has a model, then T has a model. A "model" is defined as: an interpretation I is a model for a set of formulas of a theory T,* if and only if it is true for I. One then considers a first-order theory adequate for the real numbers, R. A constant c is added to r and a denumerable infinity of axioms: there exists an x, such that x=c; c>0; c<1/2; c<1/3; … etc. The new system is called R.* By the compactness theorem, every finite subset of the axioms of R* has a model, so R* has a model. From this it is concluded that there is an object in R* which is greater than 0, but less than* any positive real number, and this, allegedly, is an infinitesimal. (Or, how about arguing that this is a counter-example to the compactness theorem?)

Or is it? One mathematical logician, Geoffrey Hunter, "Is Consistency Enough for Existence in Mathematics?" *Analysis*, vol. 48, 1988, pp. 3-5, argued that there are no unique non-standard models, and the models are not isomorphic to each other, so the set of infinitesimals is not well-defined. But, did anybody care?

However, there are other problems: "But was that existential proposition (that infinitesimals exist) established *simply* by establishing consistency? I think not. At a crucial point you have to show that every finite subset of the axioms of R* *has a model*. Or else somewhere in proving the Compactness Theorem you have to appeal to a previously proven theorem that certain sets of formulas *have models*. That is, at some point you have to establish that a *consistent* set of formulas *applies* to something." (p. 5) I think another counter-argument, from a finitist position, is to put them to proof that every – and that is at least a denumerable infinity of subsets of the axioms of R* have models – and that supertask cannot be performed.

Hunter goes on to point out that there are technical results challenging the idea that formal consistency entails having a model: there can be single sentences which are consistent, but not satisfiable due to Gödel's incompleteness theorem: L. Henkin, "Some Interconnections between Modern Algebra and Mathematical Logic," *Transactions of the American Mathematical Society*, vol. 74, 1953, pp. 410-427, at p. 425. So, Robinson has not succeeded in proving that his infinitesimals exist.

There are doubts expressed by a minority, well, by Antonio Moreno, "The Calculus and Infinitesimals: A Philosophical Reflection," *Nature and System*, vol. 1, 1979, pp. 189-201, about the cogency of non-standard analysis. Consider a point (x_0, y_0) on the curve $y=x^2$. Let dx be non-zero, but a positive or negative infinitesimal, and dy the infinitesimal change in y. The slope of a tangent to the curve at (x_0, y_0), is defined as:

Slope at (x_0, y_0)= the real number infinitely close to dy/dx.

Thus, $dy/dx=((x_0+dx)^2 - x_0^2)/dx = 2x_0+dx$. This is a hyper-real number, but since dx is infinitesimal, $2x_0+dx$, is infinitely close to $2x_0$, which is real. Therefore the slope at (x_0, y_0), is $2x_0$, the standard result. That looks good, only there is no real number infinitely close to dy/dx, which is used in the definition of the slope at the point, because real numbers simply don't get infinitely close in this non-standard sense, to anything, Moreno argues. If the slope is defined as the real part of the hyper-real number, that is consistent, and works fine. But, there was no need for all of the logical machinery to do this trick, and Leibniz could have got away with the same thing, simply by saying that the slope was the real part of the "number" and the infinitesimals were *sui generis*, inconsistencies that could be ignored by arbitrary definition.

Some further skeptical arguments are given by H. Slater, *Logic Reformed*, (Peter Lang, Bern, 2002), pp. 171-177. Slater also has an attack on Weierstrass' definition of the derivative, at least from a finitist position, because it presupposes the notion of a real number. W. Zhu (et al.), "New Berkeley Paradox in the Theory of Limits," *Kybernetes*, vol. 37, no. 3/ 4, 2008, pp. 474-481, go further, and gives an argument purporting to show that even the limit account of Weierstrass still does not escape Berkeley's paradox.

CONCLUSION

Infinity in mathematics, logic and physics is an unjustified bucket of shit that needs to be emptied, and emptied soon. Much like everything else in this fetid post-postmodern shithole.[335]

---

[335] https://www.youtube.com/watch?v=bqYpxXQYYT4&list=RDbqYpxXQYYT4&start_radio=1&t=28:https://www.youtube.com/watch?v=yjHOc2-l1As&index=8&list=RDbqYpxXQYYT4.

CONCLUSION

And, a general conclusion for this book as a whole? Buy the next one!

www.ingramcontent.com/pod-product-compliance
Lightning Source LLC
Chambersburg PA
CBHW032054080426
42733CB00006B/266